Monographs

Series Editor: U. Veronesi

The European School of Oncology gratefully acknowledges sponsorship for the production of this monograph received from Biofield Corp.

J. M. Dixon (Ed.)

Electropotentials in the Clinical Assessment of Breast Neoplasia

With 24 Figures and 22 Tables

Springer

J. Michael Dixon

Department of Surgery
The University of Edinburgh
Royal Infirmary
Lauriston Place
Edinburgh EH3 9YW, U.K.

ISBN-13:978-3-642-79996-9   e-ISBN-13:978-3-642-79994-5
DOI: 10.1007/978-3-642-79994-5

Die Deutsche Bibliothek – CIP-Einheitsaufnahme
Electropotentials in the clinical assessment of breast neoplasia: with 22 tables / J. M. Dixon (ed.). - Berlin; Heidelberg;
New York; Barcelona; Budapest; Hong Kong; London; Milan; Paris; Santa Clara; Singapore; Tokyo: Springer, 1996
 (Monographs / European School of Oncology)
 ISBN-13:978-3-642-79996-9
NE: Dixon, J. M. [Hrsg.]

Typesetting: Camera ready by editor

SPIN: 10483064   19/3133 - 5 4 3 2 1 0 — Printed on acid-free paper

Foreword

The European School of Oncology came into existence to respond to a need for information, education and training in the field of the diagnosis and treatment of cancer. There are two main reasons why such an initiative was considered necessary. Firstly, the teaching of oncology requires a rigorously multidisciplinary approach which is difficult for the Universities to put into practice since their system is mainly disciplinary orientated. Secondly, the rate of technological development that impinges on the diagnosis and treatment of cancer has been so rapid that it is not an easy task for medical faculties to adapt their curricula flexibly.

With its residential courses for organ pathologies and the seminars on new techniques (laser, monoclonal antibodies, imaging techniques etc.) or on the principal therapeutic controversies (conservative or mutilating surgery, primary or adjuvant chemotherapy, radiotherapy alone or integrated), it is the ambition of the European School of Oncology to fill a cultural and scientific gap and, thereby, create a bridge between the University and Industry and between these two and daily medical practice.

One of the more recent initiatives of ESO has been the institution of permanent study groups, also called task forces, where a limited number of leading experts are invited to meet once a year with the aim of defining the state of the art and possibly reaching a consensus on future developments in specific fields of oncology.

The ESO Monograph series was designed with the specific purpose of disseminating the results of these study group meetings, and providing concise and updated reviews of the topic discussed.

It was decided to keep the layout relatively simple, in order to restrict the costs and make the monographs available in the shortest possible time, thus overcoming a common problem in medical literature: that of the material being outdated even before publication.

Umberto Veronesi
Chairman Scientific Committee
European School of Oncology

Contents

Introduction

J. Michael Dixon[*]

Department of Surgery, The University of Edinburgh, Royal Infirmary, Lauriston Place, Edinburgh EH3 9YW, United Kingdom

Success in the treatment of bacterial infections followed identification of differences between bacterial and human cells. Although cancer cells differ biologically from cells of the host from which they arise, it has proved difficult to exploit these biological differences to improve diagnosis and treatment of solid malignancies. It has been known for some time that there are differences in the electrical potential of normal and malignant cells. With advances in electrode technology and the use of microprocessors it is now possible to record direct current over the skin surface of organs such as the breast. Early results indicate that it is possible to differentiate between benign and malignant breast masses and furthermore, it appears possible to identify breast abnormalities in asymptomatic women. This represents one of the few occasions where exploration of biologically recognised differences between normal and malignant cells has resulted in the development of a technique to differentiate normal and malignant tissues *in vivo*. As preliminary reports measuring skin potentials have indicated that this technique could have considerable clinical impact on the diagnosis and management of breast cancer, it was felt appropriate that the background scientific information and the currently available clinical data should be summarised and presented in the form of a European School of Oncology Monograph. Individuals involved in the initial development and clinical testing of this technique met in June of 1994 under the auspices of the European School of Oncology and this book is the consequence of that meeting.

[*] J. Michael Dixon acknowledges the support of the Cancer Research Campaign of the United Kingdom

Underlying Mechanisms Involved in Surface Electrical Potential Measurements for the Diagnosis of Breast Cancer: An Electrophysiological Approach to Cancer Diagnosis

Richard J. Davies

Professor of Surgery, New Jersey Medical School (UMDNJ); Chairman of Surgery, Hackensack Medical Center, 30 Prospect Avenue, Hackensack, New Jersey 07601, U.S.A.

This chapter explores at a basic level the interface between breast cancer diagnosis, cellular proliferation, cancer biology, epithelial ion transport and electrophysiology. It is not intended for the epitheliologist who wishes to understand more about epithelial biology, but rather for the epitheliologist who wishes to know how his discipline may interface with other, apparently unrelated ones.

More than 30 years ago C.P. Snow acknowledged the existence of two cultures, science and the arts or humanities [1]. Furthermore, he urged that in order to make progress these two cultures must learn to understand each other. Not only has this not happened, but the cultures of Science and Medicine themselves have become balkanized with clinicians and basic scientists failing to communicate because of the development of jargon within disciplines, a reductionist approach to science, and an exploding volume of knowledge. Sometimes, to make progress, it is necessary to think laterally and to explore the interface between traditionally narrow and isolated disciplines. Indeed, as Snow stated, "The clashing points of two subjects, two disciplines, two cultures - of two galaxies, as far as that goes - ought to produce creative chances" [1].

The Biophysical Breast Examination (BBE) is a compilation of such lateral thought and interface science, whereby observations at an ionic level have been translated into a new approach to breast cancer diagnosis.

The Basis of Cell Membrane Potential

Cell membranes are semi-permeable lipid-protein bilayers that behave as leaky electrical capacitors. A typical cell membrane is only 7 nm thick and has ions asymmetrically distributed across it. These ionic gradients are maintained in living cells by pumps. As ions tend to diffuse from a higher concentration to a lower one, the concentration gradient across the membrane results in an electrical potential (V_m), which in a typical cell is about 70 mV. The electrical field (voltage/thickness) is therefore substantial at about 100 kV/cm. This electrical field influences both ionic transport and carrier-mediated transport involving electrically charged ions. Any change in ionic transport and permeability characteristics of the cell membrane can alter the electrical field. The interrelationship of ionic transport and electrical field is thus complex [2].

In order to illustrate the genesis of an electrical potential across a cell membrane, one can consider the distribution and contribution to the electrical potential of a single permeant ion such as potassium (K^+). All cell membranes are semipermeable, which means that they are permeable to some ions but not others. K^+ is the predominant intracellular cation. If the cell membrane is semi-permeable to K^+ but not to Cl^-, then K^+ will diffuse across the cell membrane down its concentration gradient, resulting in a positive electrical potential on the outside of the cell and a negative charge on the inside of the cell membrane. This will continue until an electrical potential of about 61 mV is reached (the Nernst potential) at which further movement of potassium out of the cell (flux) is opposed because of the inside negative electrical charge. In contrast, in this situation sodium (Na+), which is the predominant extracellular cation, will tend to flow down its electrical and chemical gradient and enter the cell. The sodium-potassium ATPase pump (Na/K ATPase) maintains the chemical gradient by pumping sodium out of the cell and pumping potassium in, against their respective gradients. Without these gradients, and the pumps to maintain them, there would be no cell membrane electrical potential. There may also be a direct contribution of a number of electrogenic pumps such as the Na/K ATPase, Ca2+ or H+ pumps to cell membrane potential. These three pumps would be expected to hyperpolarize the cell membrane, i.e., make the outside of the cell more positive, because they result in a net movement of cations out of the cell [2].

Cell membranes have different permeabilities and the concentrations of intracellular and extracellular ions differ. As such they make different contributions to the cell membrane electrical potential. This may be summed up in the Goldman-Hodgkin-Katz equation:

$$V_m = -\frac{RT}{F}\ln\left(\frac{P_K[K]_i + P_{Na}[Na]_i + P_{Cl}[Cl]_o}{P_K[K]_o + P_{Na}[Na]_o + P_{Cl}[Cl]_i}\right)$$

where P_K, P_{Na} and P_{Cl} are the electrodiffusive permeability coefficients, [K], [Na] and [Cl] refer to the concentration of the given ion, the subscript i and o refers to intracellular and extracellular, respectively, RT and F refer to the gas constant, absolute temperature and the Faraday constant, respectively [2].

Because chloride is passively distributed and does not contribute to the cell membrane potential, the GHK equation may be modified to include the contribution of the Na/K ATPase pump as follows:

$$V_m = -\frac{RT}{F}\ln\left(\frac{rP_K[K]_i + P_{Na}[Na]_i}{rP_K[K]_o + P_{Na}[Na]_o}\right)$$

where r is the coupling ratio of the pump (J_{Na}/J_K) [2]. It has been estimated that the electrogenic contribution of the pump is about 10 mV [3], although the indirect contribution is much greater because of its role in maintaining a concentration gradient required to produce the membrane diffusion potential.

Cell Membrane Depolarization

As may be deduced from the formulas above, loss of the cell membrane electropotential (depolarization) may occur in one of three ways [2]:
- Change in the concentration of the permeant ions in the cytoplasm or extracellular space.
- Changes in the permeability of the cell membrane.
- Changes in the transport of electrogenic pumps.

All the above changes have been observed in proliferating cells, mitogenesis and malignant transformation and the changes in relation to proliferation and cancer are discussed below.

Cell Membrane Depolarization and Proliferation

In 1971 Hülser and Frank observed that the addition of serum to quiescent fibroblasts resulted in rapid cell membrane depolarization [4]. The addition of serum to cultured cells is required to stimulate cell division and growth. This observation therefore led to the notion that cell membrane depolarization was an early event associated with cell division. Further observations in other cell lines demonstrated that depolarization induced by growth factors is biphasic [5]. Subsequent experiments, however, showed that epidermal growth factor (EGF) stimulated cell division without depolarization [6], and other authors have reported that growing or quiescent cells in culture may have similar cell membrane voltages. Reuss et al. reported that cultured BSC-1 epithelial cells had a cell membrane voltage of -48 mV and depolarized by 5 to 20 mV following the addition of serum or EGF [7]. This was followed by repolarization in 5-10 minutes. Although this depolarization is temporally associated with Na+ influx, the influx persists after repolarization occurs [8]. This suggests that although the initial Na+ influx may result in depolarization, the increase in sodium transport does not cease once the cell membrane has been repolarized, possibly due to Na/K ATPase pump activation (Fig. 1) [9].

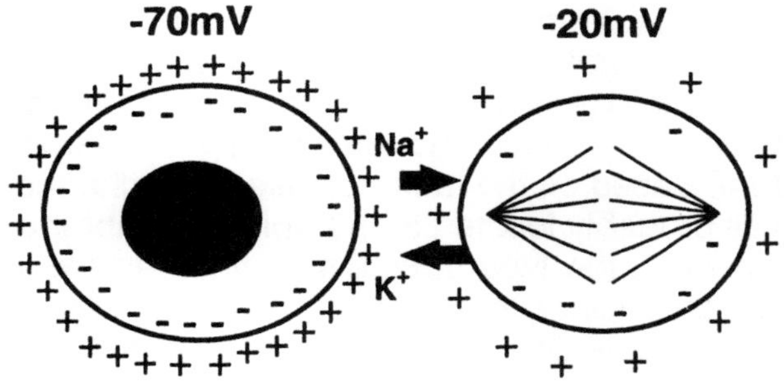

Fig. 1. Electrical depolarization during cell division or mitosis. The -70mV normal cell membrane potential is dissipated to -20mV as extracellular Na+ and intracellular K+ enter and leave the cell, respectively, down their concentration gradients. This phenomenon has been observed in a number of different cell types.

A number of studies have confirmed that proliferating cells are relatively depolarized when compared to their non-dividing or resting counterparts [10-14]. Other studies which have examined alterations in ionic fluxes, intracellular ionic composition and transport mechanisms associated with mitogenesis, have indicated that the cell membrane potential may depolarize during cell activation and proliferation [15-26].

Both intracellular Ca^{2+} (Ca^{2+}_i) and pH (pH_i) are increased by mitogen activation [7,18,20-25]. These in turn may alter the gating of various ion channels in the cell membrane, which are responsible for maintaining cell membrane voltage [7,11-19]. There is thus the potential for interaction between other intracellular messengers and cell membrane potential. Nonetheless, it does not appear that cell membrane depolarization is a prerequisite for cell division even though it is a frequently associated event in cultured cells.

None of these studies imply that cell membrane depolarization is causally related to the onset of proliferation, only that it is frequently associated with cell mitosis or activation. A number of investigators have failed to demonstrate mitogen-induced cell membrane depolarization [8, 25-26], although the mitogens selected may have "bypassed" earlier steps in the activation sequence, including cell membrane depolarization. One problem with many of the above studies is that the majority have used cultured tumour cells and therefore it may be misleading to extrapolate from these cells to non-transformed cells undergoing proliferation.

Cell Membrane Depolarization and Cancer

Studies in the 1950s and 1960s, using conventional intracellular glass microelectrodes, demonstrated that cancer cells are relatively depolarized compared with non-transformed cells [27-29]. In 1971 Cone suggested a "unified theory" of mitogenic control in which sustained cell membrane depolarization resulted in continuous cellular proliferation [30-31]. It was further postulated that malignant transformation resulted from sustained depolarization and a failure of the cell to repolarize after cell division [32]. In support of this view, the temperature-sensitive Moloney sarcoma virus, when used to infect and transform kidney cells in culture, results in cell membrane depolarization which precedes other transformation-specific events [33].

Other studies have also demonstrated cell membrane depolarization during transformation and carcinogenesis [34-36]. More recent studies from our own laboratory have shown that there is a progressive depolarization of the colonocyte cell membrane during 1,2 dimethylhydrazine (DMH)-induced colon cancer induction in CF_1 mice [37]. The V_A (apical membrane voltage) measured with intracellular microelectrodes in apparently "normal" colonic epithelium depolarized from -74.9 mV to -61.4 mV following 6 weeks of DMH treatment; to -34 mV by 20 weeks of treatment.

While epithelia normally maintain their intracellular sodium concentration within a narrow range [2,9], electronmicroprobe analysis suggests that cancer cells exhibit cytoplasmic sodium/potassium ratios that are 3 to 5 times greater than those found in their non-transformed counterparts [38,39]. These observations may explain in part the electrical depolarization observed in malignant or premalignant tissues, which could reflect the loss of K+ or Na+ gradients across the cell membrane.

In addition to cell membrane depolarization and altered intracellular ionic activity, studies have shown that there may be a decrease in electrogenic sodium transport and activation of non-electrogenic transporters during the development of epithelial malignancies [40]. These changes may affect or occur as a consequence of altered intracellular ionic composition.

Patch Clamp Studies in Mammary Epithelial Cells

Ions move across cell membranes in a number of ways including pumps which require energy, transporters in which ions are exchanged or accompany one another in order to maintain electroneutrality, and along ionic channels which have specific conductance properties. It is ionic channels which have the greatest influence on cell membrane electrical potential. The development of the patch clamp technique has enabled the examination of the conductance and electrical properties of single cells and indeed single channels. The technique uses a high impedance amplifier and microelectrode to suction part of the cell membrane onto the electrode tip. The patch is then sealed by suction onto the electrode so that the electrical properties of the channels or single channel in the patch can be examined. Variations of this technique include cell-attached, cell-detached patches, or patch-permeabilized patches to look at the electrical properties of the whole cell rather than the isolated patch. The importance of these techniques is that one can examine the electrical properties of a cell or individual ionic channel under different growth activation conditions of the cell. For example, by applying varying currents under voltage-clamped conditions one can examine the current-voltage (I/V, i.e., conductance) characteristics of an individual ionic channel. The conductance as well as the open probability of the given channel may therefore be estimated (the open probability is the chance of a given channel being open at a known cell membrane potential). A cell membrane may therefore depolarize for a number of reasons, including a decrease in the number of ionic channels, a decrease in the open probability, the replacement of high conductance channels by low conductance channels, or a change in the selectivity of a channel.

Two studies have been published that have applied the patch clamp technique to mammary epithelial cells. The first by Furuya et al. [41] used a primary culture from lactating mice and demonstrated a K^+ channel of small unitary conductance which is activated by nanomolar concentrations of Ca^{2+}. This channel had a non-linear current-voltage relation (inward rectifying) with a slope conductance of 12 pS (picoSiemens, a measure of conductance) for outward current and 36 pS for inward current. The channel was competitively blocked by Na^+ and TEA^+ (tetraethylammonium) from the cytoplasmic side, and was voltage independent. Based on the methods used, the channel was probably located in the apical membrane. However, the type of cell, and whether it was ductal or acinar, is unclear. The authors speculate that this Ca^{2+}-dependent K^+ channel may be responsible for the oscillatory, spontaneous hyperpolarizing changes in membrane potentials observed after exposure to growth factors such as EGF [41]. Other channels may downregulate or close in response to hyperpolarizing growth factor-induced changes so that the overall effect on an intact epithelium cannot be determined. However, it does suggest an ionic basis to the electrical changes associated with exposure to growth factors.

Another patch clamp study was of apical membrane K^+ channels in the MCF7 breast carcinoma cell line [42]. The principal channel identified was a 23 pS Ca^{2+}-activated K^+ channel with a linear I/V relation (used to estimate and characterize conductance of channel) and an increased open probability upon depolarization. This would tend to counter any depolarizing effects of mitogens. The channel is highly selective for K^+ over Na^+ and was not blocked by TEA^+, quinidine or apamin. It was most prevalent in rapidly cycling cells, but was not required for cell division, suggesting that it is associated with cell division rather than being causally related to it. Its occurrence was decreased in cells in which growth is inhibited by addition of phorbol 12-myristate 13 acetate (TPA) or by tamoxifen to the growth media. The incidence of this channel, 78% of patches in cells in the exponential growth phase, was reduced to 42% in cells grown in culture media containing TPA, while tamoxifen reduced the incidence from 95% to 60%. The authors suggest that this channel may be an essential part of the mechanism by which an MCF7 cell is maintained in the rapidly cycling pool during exponential growth. However, for most epithelial cells functions pertaining to cell homeostasis and growth are usually relegated to the blood-facing or basolateral membrane, while those governing transport rate are often in the apical membrane. It should be remembered that proliferating cells are usually insufficiently differentiated to have a distinct separation between transport proteins on the apical and basolateral cell membranes, which, as discussed below, may explain the reason why intact epithelial cells with dysregulated proliferation may be incapable of maintaining a transepithelial potential. Thus, it is unclear whether this channel would be directly involved in maintaining rapid cell cycling in intact breast epithelium. If this channel is physiologically significant, it would tend to hyperpolarize breast epithelium or alternatively it may appear as a homeostatic mechanism in an attempt to restore the cell membrane potential in a depolarized proliferating cell population. In either case it does appear that electrical changes occur in breast epithelial cells as a consequence of altered proliferation and may be measured at the single ionic channel level.

Transepithelial Electrical Potential

Epithelial cells line many solid organs if secretion and absorption are a part of their function. Examples include stomach, colon, prostate, endometrium, lung and breast. These are also the sites of common malignancies. All these organs absorb and secrete various ions and water. The transepithelial electrical potential has been studied in many of these organs and the potential is invariably luminal side (apical) negative in relationship to the abluminal, or bloodstream, side (basolateral). The cell membrane on the other hand is negative inside in relation to the outside of the cell. The reason for the luminal side being electrical negative is that

the permeability properties are quite different for the apical compared with the basolateral cell membrane. For example, in most epithelia the luminal or apical cell membrane is predominately permeable to sodium, which permits sodium to enter the cell down its electrical and concentration gradient. The Na/K ATPase pump then extrudes sodium into the abluminal space across the basolateral membrane and water either follows through the cell or through tight junctions between cells. This permits the absorption of Na^+ and water in the small or large intestine. Other epithelium, such as the gastric lining, has specialized apical and basolateral membranes to permit secretion of H^+ and Cl^- ions (Cl^- against its electrical gradient). Microelectrode studies in this type of epithelium illustrate that there is an electrical gradient of about 40-60 mV across the apical membrane (inside negative) and 50-100 mV across the basolateral membrane (inside negative) which results in a 10-40 mV transepithelial electrical potential [2,43]. The transepithelial electrical potential is thus derived from the sum of the apical and the basolateral potentials; i.e., V_T (transepithelial potential) is the electrical potential that exists across the epithelium and is the algebraic sum of the apical membrane (V_A) and basolateral membrane (V_{BL}) potentials of cells arranged in an epithelial sheet or inside a hollow viscus, such as the cells lining the lobules and ducts of the breast or the lumen of the bowel. Therefore $V_T = V_A + V_{BL}$ and the BBE test measures V_T at the skin surface (see Fig. 2).

Fig. 2. Electrical gradients across normal breast epithelium lining the terminal ductal lobular units of the breast (TDLUs). Because of the higher charge density across the basolateral, compared to the apical cell membrane, the basolateral voltage or potential difference V_{BL} is -100 mV vs -70 mV for the V_A (apical membrane voltage) results in a net transepithelial voltage V_T of -30mV (luminal side negative). The cell membrane is divided into distinct apical and basolateral domains by the tight junctions in order to facilitate vectorial transport. Na^+ channels on the apical or ductal luminal aspect permit "downhill" entry of Na^+ into the cell. The Na/K pump on the basolateral membrane

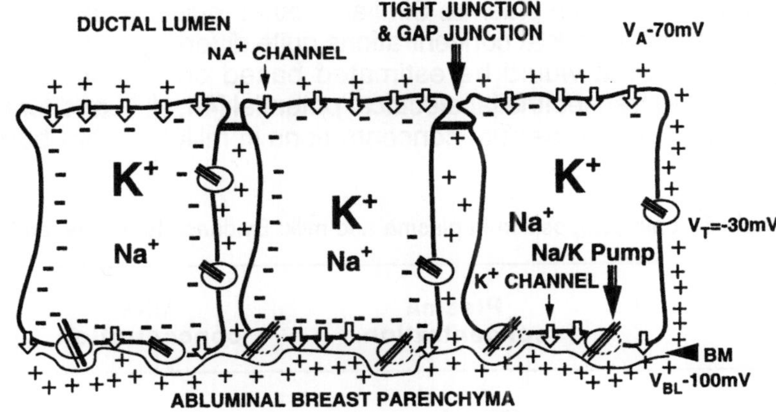

is integral to maintaining the electrochemical profile. K$^+$ channels are predominantly on the basolateral cell membrane and the ionic gradient due to K$^+$ is largely responsible for cell membrane potential. Water and solutes can cross the epithelium either between the cells (paracellular) or through the cell (transcellular). BM depicts the basement membrane on which the epithelial cells sit.

Changes in ionic concentration, individual cell membrane permeabilities or electrogenic pump activity may influence not just individual cell membrane electrical potentials, but could also affect the transepithelial electrical potential.

Transepithelial Ion Transport and Electrical Potential in Mammary Epithelia

Most studies of ion transport in breast epithelium have been performed in lactating mammals, or in cultured cell lines. The non-lactating breast maintains its epithelial structure, consisting of ducts ending in alveoli with morphology analogous to clusters of grapes. There are few studies of basal ion transport in this epithelium when it is in the non-lactating state [44]. Those studies that do exist suggest that ion transport and therefore transepithelial potential may be altered in the malignant state [45,46]. Lactation may protect against breast cancer development [47], and would be expected to increase (hyperpolarize) transepithelial electrical potential because of active ion secretion.

In the lactating breast, milk is secreted slowly at a rate of 1-2 ml milk/g tissue per day and is stored in alveoli and ducts, often for many hours, until suckling. The epithelia must therefore deal with two transport problems: secretion, which determines the initial composition of milk, and storage, during which the composition of the milk must be maintained for variable times in the presence of substantial ion gradients. The aqueous phase of milk is isotonic in relation to plasma, but has a much lower sodium content than plasma. Similarly, other ions are present in milk at concentrations quite different from what would be estimated based on measured transepithelial electrical potential [44, 48]. Although potassium concentrations in milk

are close to what would be estimated from its Nernst equilibrium potential, other large ion concentrations differ and the gradients are probably maintained by active ion transport which will affect transepithelial potential. The transepithelial electrical potential during lactation averages -35±2 mV in the mouse [49]. Table 1 summarizes the ionic composition of milk from the literature [44,48] and estimates the Nernst equilibrium concentrations if no active ion transport were involved in milk production. There are no *in vivo* studies in quiescent non-lactating breasts, so that little is known about basal transport and its contribution to transepithelial electrical potential.

The exact mechanisms of ion translocation at apical and basolateral membranes are not well established. It is clear, however, as is common in most other epithelia, that Na, K-ATPase pumps are present in the basolateral membrane of mammary epithelial cells [50] and that these provide the driving force in terms of transmembrane ion gradients. Clearly, Na+ at a milk concentration of 6.5 mM is much less than its equilibrium value of 530 mM and, therefore, either the cellular and paracellular pathways must be impermeant to Na+, Na+ must be actively absorbed, or both. Similarly, while K+ may be at transepithelial equilibrium, Cl- is half the expected value, suggesting active transport of this ion. Recognition of these disequilibrated ionic distributions has led several investigators to evaluate the ion permeability of the apical membrane. Studies by Linzell and Peaker [51,52] suggested that Na+ and K+ are passively distributed across the apical membrane since the molar ratio of Na/K in milk is similar to that in cytoplasm. More recent studies in the mouse have suggested that the apical membrane contains channels for both Na+ and K+ and for Cl- and that the ionic distribution is such that both Na+ and K+ cannot be in equilibrium

Table 1. Ionic composition in plasma and milk: Evidence for active ion transport

Ion	Plasma concentration	Milk concentration	Nernst equilibrium concentrations
Sodium	138 mM	6.5 mM	517 mM
Potassium	4.6 mM	14.6 mM	17 mM
Chloride	107 mM	12.1 mM	29 mM
Calcium	2.25 mM	4.0 mM	8 mM

across the apical membrane [49]. In other recent studies where the ionic composition of the lumen was altered in goat mammary gland, a notable feature was the inability of the apical membrane to discriminate between Na+ and K+ [53], leading the investigators to conclude that the apical membrane permeabilities to Na+ and K+ are virtually identical either because of similar numbers of channels for each ion or because of the presence of non-specific channels in the apical membrane. As was discussed above, patch clamp studies in cultured mammary epithelial cells report that apical K+ channels are highly specific for K+ over Na+ [41,54]. However, in isolated colonic crypts, a tissue that may have similar features to the breast epithelium, the apical membrane displays both highly specific K+ channels and non-specific channels that are activated by cAMP [55]. Therefore it seems that our understanding of the mechanisms underlying regulation of mammary ion transport at the channel level is still very much in its infancy and no firm conclusions can yet be drawn.

Altered Ion Transport in Breast Cysts and Breast Cancer

Gross cystic disease of the breast occurs in about 7% of the female population and consists of fluid-filled cysts of more than 3 mm developing in the terminal ductal lobular units (TDLU). Apocrine metaplasia, which can develop in the TDLU epithelium, appears to form into apocrine microcysts which may subsequently develop into gross cysts. Although breast cysts are considered to be benign, those patients with cysts lined by apocrine epithelium appear more likely to develop breast cancer than patients whose cysts are lined by less active flattened epithelium. Apocrine breast cysts have been classified as type I cysts and contain high concentrations of potassium and low concentrations of sodium relative to plasma [56]. The Na+/K+ ratio in these cysts is ≤ 3.0. On the other hand simple cysts, designated type II cysts, contain electrolytes closer to plasma concentrations with Na+/K+ ratios >3.
There have been a number of studies examining growth factor levels in different types of cysts [57] and attempts have been made to correlate the presence or absence of given growth factors with the proliferative activity of the epithelium. Little attention has been given to the significance of the altered electrolyte content in these epithelial-lined structures. If cyst content reflects electrolyte composition of surrounding breast tissue and interstitial fluid, as has been suggested, then one would expect to find relative depolarization of the epithelium if the primary source of the electrical potential is a K+ diffusion potential similar to other types of epithelium [2,58]. This is thus supportive evidence for epithelial depolarization in breast epithelium.

Transepithelial Electrical Depolarization and Cancer

A number of experimental studies have demonstrated that transepithelial depolarization is an early feature of the premalignant state [59, 40]. These studies have been primarily performed *in vitro* and in experimental models of large bowel cancer. More recent studies have also demonstrated that in direct measurements of electrical potential in biopsied breast specimens, electrical depolarization was found to be associated with malignancy [60]. Recently alterations in impedance have been found to be associated with both premalignant and cancerous states [61,62]. Impedance is related to voltage or potential difference through Ohm's law; $V = IR$. Therefore it is likely that the electrical potential may be altered as a consequence of development of cancer.
Our laboratory has demonstrated that transepithelial depolarization is a specific event associated with colonic carcinogenesis in CF_1 mice [59]. The more susceptible site, the distal colon, underwent an approximately 30% decrease in transepithelial potential (V_T) after only 4 weeks of carcinogen treatment. This was before histological changes developed [63]. A non-specific cytotoxic agent (5-fluorouracil) administered over the same period did not cause a reduction in V_T in the same model [59]. The reduction in V_T was confirmed in a subsequent study where up to a 60% reduction was observed after carcinogen treatment [40]. Both these studies were performed *in vitro* and some "run-down" in V_T is invariably noted under these conditions. We have noted that, al-

though V_T is invariably higher when measured *in vivo*, the "premalignant" colonic epithelium is usually depolarized when compared to normal colon [Davies, unpublished]. It was also noted that the murine strain DBA/2, which is relatively resistant to carcinogen-induced colon cancer, undergoes an almost 30% reduction in V_T following carcinogen treatment [40], suggesting that depolarization may be an associated rather than a causative event in the carcinogenic process.

Field Carcinogenesis and Proliferative Breast Disease

Histological Evidence

Field cancerization was originally described by Slaughter in head and neck malignancies [64]. All surgical specimens examined in 783 cases resected for malignancy contained epithelial hyperplasia and keratinization of the tumour margins with 11.2% of patients having multiple tumours. The authors suggested that the mucosa was "condemned" by a process they called "field cancerization" in which it was postulated that the entire mucosa had subtle biochemical, molecular or genetic alterations associated with carcinogen exposure and should be considered as "precancerous". Since these seminal observations there has been a flurry of interest in precancerous biomarkers associated with field cancerization which may precede or are associated with the development of cancer [65].

In 1941, several years before these observations, Sir Robert Muir stated that "in a large proportion of cases of carcinoma of the breast, the stages of evolution of malignancy within ducts and acini cannot be followed and in some of these there is evidence that malignancy arises *de novo* without the occurrence of preliminary hyperplastic changes" [66]. Gallager and Martin in a careful pathological study observed that some form of epithelial hyperplasia could be observed in 100% of "cancer-associated breasts" [67].

Histologically there is a considerable body of evidence that breast epithelium derived from patients with breast cancer is "condemned" in the sense that it is precancerous. This is suggested by the development of second breast cancers in women treated for the disease with an incidence of 1-2% each year in the contra-lateral breast, and by the 44-75% incidence of multifocality (i.e., second invasive or *in-situ* carcinomas being present in the same breast) [68-72]. Simpson identified an increased prevalence of epithelial hyperplasia or epitheliosis in premenopausal women with breast cancer [73]. Although epithelial hyperplasia occurs normally in about 20% of premenopausal women, it is found in the adjacent "normal" breast with an incidence of about 70% in premenopausal women who have breast cancer [73]. It is of note that following the menopause there is a decrease in the frequency with which epithelial hyperplasia is found in the normal premenopausal breast although later in life it begins to increase and reaches about 50% in the normal breasts of women by age 80-90. The increased frequency of epithelial hyperplasia above control levels found in the premenopausal woman with breast cancer was not observed in postmenopausal women with the disease.

More recent studies have demonstrated that fine-needle aspiration of the breast in women with a strong family history of breast cancer who had not yet developed the disease revealed proliferative changes in 35% of these clinically normal females compared with only 13% of controls [74]. This supports the concept of proliferative instability in women with breast epithelium "at-risk" of developing breast cancer.

It has been estimated that the minority of women with benign proliferative breast disease go on to develop breast cancer [75]; this number rarely exceeds 10%. Rosen recently stated that "Currently there are no procedures or tests available for determining which specific proliferative changes are precursor lesions committed to, capable of or susceptible to evolving into carcinoma" [75]. Although histologically there appears to be a continuum of changes from simple epithelial hyperplasia, through more severe forms of hyperplasia with atypia, to ductal carcinoma *in-situ* and invasive carcinoma [76], there are no histological means of determining which lesions are committed to evolve into cancer (obligate precursor lesions) and these lesions are merely marker lesions which may indicate proliferative instability but not epithelium committed to transformation. Nevertheless, from a histological standpoint a field effect of proliferative instability appears to be associated with the breast epithelium "at-risk" of cancer development.

Cell Kinetic Evidence for Proliferative Instability

In many epithelial malignancies, studies of cell cycle kinetics using mitotic index or labelling index support the notion that cancers develop on a background of disordered proliferation [77-79]. One advantage of these studies, over the histological evidence cited above, is that they attempt to measure the rate of cell turnover and the length of cell cycle as a measure of cell proliferation kinetics. They also take into account the location of the proliferating cells in colonic crypts and any expansion of the proliferative compartment [77-79]. The major disadvantages are variability within individual patients and variability between observers and laboratories, and assumptions that *in vitro* measurements, during which the proliferative activity may rapidly drift toward a more basal level, reflect the situation *in vivo*. Variability may also occur because there are localized islands of proliferative instability in the breast as in the colon, where there are aberrant crypt foci [80]. This makes estimation of labelling indices subject to marked variability and explains some of the variation reported in studies of breast epithelium [81-88].

Kinetic evidence for upregulated proliferation in the breast, as opposed to dysregulated proliferation, was suggested from non-controlled studies [81], but more recent studies do not confirm a generalized upregulation in proliferation [82]. Cellular proliferation in the normal resting (non-lactating) breast remains a topic of ongoing study. Most experiments have been performed *in vitro* using "normal" breast tissue adjacent to benign or malignant breast disease or tissue obtained from reduction mammoplasty specimens. Van Bogaert has measured thymidine labelling indices (LI) in intralobular ductules, terminal ducts and large ducts and found the LI to vary between 0.4% and 1.0% [83, 84]. Meyer measured the LI in morphologically normal terminal ducts from 104 specimens of tissue adjacent to benign breast disease [85]. The LI varied with the patient's menstrual cycle [86]. It has also been demonstrated that the LI appeared to inversely correlate with the age of the patient [86]. This observation has been

confirmed by Potten [87] and in a smaller study it was shown that the LI of the TDLUs was higher in premenopausal than in postmenopausal women [88].

Flaxman and Lasfargues performed an early study of so-called "normal" epithelium in non-neoplastic ducts in areas remote from the malignancy [81]. They found that the mean LI was 3-4%, which is 4 to 5 times higher than reported in breast epithelium from patients with benign breast disease [82-88]. This finding raises the possibility that there is an upregulation in proliferation occurring throughout the breast in patients with breast cancer similar to the increase in mucosal proliferation observed in patients who develop adenomas or colon cancer [77-79]. This generalized upregulation in proliferation may explain the finding with the Biofield device that size and depth of a lesion do not adversely affect the sensitivity of the test [personal communication M. Faupel and 89], i.e., depolarization occurs throughout the involved quadrant of the breast because the surrounding breast tissue is involved in this electrical depolarization (field depolarization). However, more recent studies do not confirm an increase in proliferation in the TDLUs of cancer-associated breasts [82]. One study has shown that the labelling index was higher in the "normal" terminal ducts and alveoli close to a breast cancer but not in those further away [88]. The study reported that the labelling index in the ductal epithelial cells of TDLUs close to the tumour was 0.74 ± 0.88% (mean ± SD) compared to 0.50 ± 0.41% in the TDLUs distant from the tumour. Similarly, the labelling index was 0.71 ± 0.77% compared to 0.61 ± 0.60% in the lobules close to and distant from the breast cancer. Although this study is cited as a negative study with regard to there being no upregulation in proliferation in the region around a developing cancer, the 50% higher LI in TDLUs near to the tumour is in keeping with this hypothesis, even though statistical difference was not reached because of small sample size (n=7) and a large standard deviation. Another study by Allan has failed to observe proliferative upregulation around a breast cancer but does note that the apoptotic-proliferative ratio is reduced in the TDLUs surrounding a tumour [82]. It was noted that LI and mitotic index (MI) were both reduced in the TDLUs of cancer-associated breasts compared with those from patients with benign breast disease. The LI was only 2.1 ± 0.4% (mean ± SEM) in cancer-associated breasts compared to 2.3 ± 0.2 and 2.4 ± 0.3 in fibroadenoma-associated and fibrocystic-associated breasts, respectively, after adjustment for age and menstrual cycle. Despite the normal proliferative indices this study showed higher apoptotic to mitotic ratios in cancer-associated and fibrocystic-associated breasts than normal fibroadenoma-associated breasts. This suggests that there is an imbalance between the number of cells dying and the number of new cells being formed in TDLUs in cancer-associated breasts, although fibroadenoma-associated breast epithelium may in fact not be normal. Therefore these studies should be interpreted with caution. The effect of this phenomenon on the electrical properties of any epithelium has never been examined. One significant problem with studies of epithelial proliferation has been marked heterogeneity in the measurements both within and between individual studies [90]. This may be because islands of prolifera-

tive upregulation develop in the epithelium "at risk" for cancer; this certainly occurs in the colon with the development of aberrant crypt foci and polyps [80]. Until more definite evidence is obtained that proliferation is upregulated, it may be more correct to state that proliferation is dysregulated in the zone around a developing cancer. This zone or field of dysregulated proliferation results in a penumbra of changes that extends into the surrounding breast parenchyma and epithelium. Preliminary *in vivo* studies using C^{11}-labelled thymidine and PET scanning suggest that there may be a proliferative penumbra around breast cancers [personal communication W. Dooley]. The size of this penumbra has not been determined, but other investigators have noted other changes in breast parenchyma and TDLUs which are summarized in Table 2 and suggest the possibility of a field effect of the proliferation, biochemistry, molecular biology or genetics of cancer-associated breasts.

Table 2. Field changes associated with breast cancer

Type of Cellular Alteration	Reference
1. Abnormal epithelial microvillae	Spring-Mills [91]
2. Proliferative upregulation	Flaxman [81]
3. Increased variability of cell surface antigens	Courtney [92] & Ceriani [93]
4. Increased C-11 thymidine incorporation *in vivo*	Personal communication [94]
5. Increased impedance [1]	Morimoto [62]
6. Electrical depolarization of the breast [2]	Marino [60]
7. Reduced alpha B4 Integrin expression	Jones [95]
8. Apoptosis reduced but no proliferative upregulation [3]	Allan [82]
9. Increased pre-ovulatory hyperthermia	Simpson [96,97]
10. Increased 16-α hydroxylation of estrone [4]	Telang [98] & Osborne [99]
11. Increased pentose phosphate shunt activity	Weber [100] & McDermott [101]
12. Increased EGFR and c-*myc*	Whittaker [102]
13. Increased angiogenesis	Jensen [103]
14. Increased intralobular fetal fibroblasts	Schor [104]

1 Both the resistance of the breast mass itself (R_i) and the surrounding tissue (R_e) were increased if the mass was demonstrated to be a cancer, suggesting that the surrounding tissue was also affected
2 Electrical depolarization was measured in breast biopsies taken from patients with breast cancer compared with benign masses *in vitro*. It was not clear whether these measurements were made in the apparently "normal" tissue or in tumour itself
3 It is not clear whether "normal" samples were taken immediately adjacent to the tumour or some distance away. Labelling indices were less in cancer or fibrocystic-associated breasts than in normal tissue adjacent to fibroadenomas
4 This oestrogen metabolite may increase proliferative activity in breast TDLUs

Underlying Mechanisms

It has been explained how cells develop and maintain an ionic gradient which results in an electrical potential gradient across the cell membrane. Epithelial cells in the breast are arranged to line ducts or lobules and have specialized apical and basolateral domains to facilitate basal absorption, secretion and milk production during lactation. Because these individual membranes have different permeabilities and transport functions, they have different electrical potentials in relationship to each other and therefore result in a transepithelial electrical potential. Since cells undergoing proliferation, with different metabolic characteristics or cell surface characteristics may become electrically depolarized, the transepithelial electrical potential of the surrounding breast ducts and lobules also becomes depolarized. Proliferation is increased in cancer cells and proliferation becomes dysregulated in surrounding breast terminal ductal lobular units. Field cancerization results in field depolarization, which extends in a penumbra measurable at the skin surface.

Early in the process of carcinogenesis gap junction-mediated intercellular communication becomes impaired [105]. The loss of gap junction integrity leads to loss of both cell-to-cell coupling and cell adhesion [106]. Loss of electrical coupling between cells may result in foci of depolarization, which may in turn explain the increased differential found in "cancer-associated" and "at-risk" breasts. The differential is the difference between the maximum and the minimum transepithelial potential measured at the skin surface. The larger differential found in the "cancer-associated" breast is a reflection of the heterogeneity in depolarization (see Fig. 3). This may be expected given the variability in

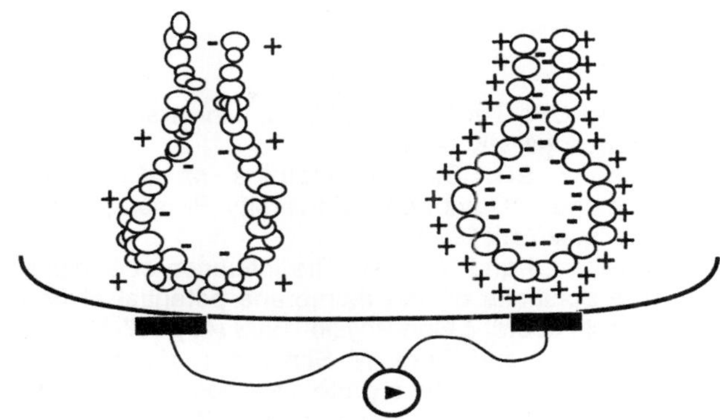

Fig. 3. Hyperplastic terminal ductal lobular unit (TDLU) on the left or cells undergoing dysregulated proliferation within a TDLU associated with a nearby breast carcinoma. The TDLU has undergone transepithelial depolarization as a consequence. The adjacent normal TDLU, in a region of the breast without the breast cancer has a normal transepithelial potential profile. An electrode or sensor placed on the skin surface shows a significant electrical differential between these two regions. The high differential indicates the presence of a nearby malignancy.

proliferative upregulation found in the breast [90]. Therefore, both depolarization relative to the opposite breast and increased differential appear to be features of premalignant breast epithelium.

The ability to measure this altered transepithelial electrical potential using skin electrodes opens a new approach to breast cancer diagnosis using non-invasive means, and does not represent a standard imaging approach. The BBE employs a biophysical property of epithelium "at-risk" of breast cancer to assist with breast cancer diagnosis. It has profound implications not just for diagnosis but also for screening and possibly identifying those at risk of this common disease.

REFERENCES

1   Snow CP: The Two Cultures. University Press, Cambridge, UK 1959

2   Schultz SG: Basic Principles of Membrane Transport. Cambridge University Press, London and New York 1980

3   Mullins LJ and Noda K: The influence of sodium-free solutions on the membrane potential of frog muscle fibers. J Gen Physiol 1963 (47):117-132

4   Hülser DF and Frank W: Stimulation of embryonic rat cell in culture by a protein fraction isolated from fetal calf serum. Z Naturforsch 1971 (26b):1045-1048

5   Moolenaar WH, De Laat SW, Van der Saag PT: Serum triggers a sequence of rapid ionic conductance changes in quiescent membrane cells. Nature (London) 1979 (279):721-723

6   Moolenaar WH, De Laat SW, Mummery CL, Van der Saag PT: Na+/H+ exchange in the action of growth factors. In: Boynton AL, McKeehan WL, Whitfield JF (eds) Ions Cell Proliferation and Cancer. Academic Press, New York1982, pp 151-162

7   Reuss, L, Cassel D, Rothenberg P, Whiteley B, Mancuso D, Glaser L: Mitogens and ion fluxes. In: Mandel LJ and Benos DJ (eds) Current Topics in Membranes and Transport. The role of Membranes in Cell Growth and Differentiation. Vol 27. Academic Press, Orlando 1986 pp 3-54

8   Rothenberg P, Reuss L, Glaser L: Serum and epidermal growth factor transiently depolarize quiescent BSC-1 epithelial cells. Proc Natl Acad Sci USA 1982 (79):7783-7787

9   Schultz SG: Homocellular regulatory mechanisms in sodium-transporting epithelia: avoidance of extinction by "flush-through". Am J Physiol 1981 (241):F579-F590

10   Sachs, HG, Stambrook PH, Ebert JD: Changes in membrane potential during the cell cycle. Exp Cell Res 1974 (83):362-366

11   Moolenaar WH, de Laat SW, van Der Saag PT: Serum triggers a sequence of rapid ionic conductance changes in quiescent neuroblastoma cells. Nature (London)1979 (279):721-723

12   Kiefer H, Blume AJ, Kaback HR: Membrane potential changes during mitogenic stimulation of mouse spleen lymphocytes. Proc Natl Acad Sci.USA 1980 (77):2200-2204

13   Moolenaar WH, Mummery CL, Van Der Saag PTT, De Laat SW: Rapid ionic events and the initiation of growth in serum-stimulated neuroblastoma cells. Cell 1981 (23):789-798

14   Chapman LM and Wondergen R: Transmembrane potential and intracellular potassium ion activity in fetal and maternal liver. J Cell Physiol 1984 (121):7-12

15   Felber SM and Branad MD: Concanavilin A causes an increase in sodium permeability and intracellular sodium content of pig lymphocytes. Biochem J 1983 (210):893-897

16   O'Donnell ME and Villereal ML: Membrane potential and sodium flux in neuroblastoma X glioma hybrid cells: Effects of amiloride and serum. J Cell Physiol 1982 (113):405-412

17   Leffert HL and Koch KS: Ionic events at the membrane initiate rat liver regeneration. Ann NY Acad Sci 1980 (339):201-215

18   Villereal ML: Sodium fluxes in human fibroblasts: Effects of serum Ca2+ and amiloride. J Cell Physiol 1981 (107):359-369

19   Fehlmann M, Canivet B, Freychet P: Epidermal growth factor stimulates monovalent cation transport in isolated rat hepatocytes. Biochem Biophys Res Commun 1981 (100):254-260

20   Moolenaar WH, Tertoolen LGJ, Tsien RY, Van Der Saag RT, De Laat SW: Na+/H+ exchange: Cytoplasmic pH and the action of growth factors in human fibroblasts. Nature (London) 1983 (304):645-648

21   Paris S and Pouyssegur J: Biochemical characterization of the amiloride-sensitive Na+/H+ antiport in Chinese hamster lung fibroblasts. J Biol Chem 1983 (258):3503-3508

22   Paris S and Pouyssegur J: Growth factors activate the Na+/H+ exchange system in quiescent fibroblasts by increasing its affinity for intracellular H+. J Biol Chem 1984 (259):10980-10994

23   Pouyssegur J, Chambard JC, Franchi A, Paris S, van Obberghen-Schilling E: Growth factor activation of an amiloride-sensitive Na+/H+ exchange system in quiescent fibroblasts: Coupling to ribosomal protein S6 phosphorylation. Proc Natl Acad Sci USA 1982 (79):3935-3939

24   Pouyssegur J, Sardet C, Franchi A, L'Allemain G, Paris S: A specific mutation abolishing Na+/H+ antiport activity in hamster fibroblasts precludes growth at neutral and acidic pH. Proc Natl Acad Sci USA 1984 (81):4833-4837

25   Moolenaar WH, Tertoolen LG, DeLaat SW: The regulation of cytoplasmic pH in human fibroblasts. J Biol Chem 1984 (259):7563-7569

26   Deutsch C and Price M: Role of extracellular Na+ and K+ in lymphocyte activation. J Cell Physiol 1982 (113):73-79

27   Schaefer H and Schanne O: Membranpotentiale von Einzelzellen in Gewebekulturen. Naturwissenschaften 1956 (43):445

28   Tokuoka S and Morioka H: The membrane potential of the human cancer and related cells. Gann 1957 (48):353-354

29   Balitsky KP and Shuba EP: Resting potential of malignant tumour cells. Acta Union Int Cancer 1964 (20):1391-1393

30   Cone CD: Unified theory on the basic mechanism of normal mitotic control and oncogenesis. J Theor Biol 1971 (30):151-181

31   Cone CD and Cone CM: Induction of mitosis in mature neurons in central nervous system by sustained depolarization Science 1976 (192):155-158

32   Cone CD: The Role of surface transmembrane potential in normal and malignant mitogenesis. Ann NY Acad Sci 1974 (238):420-435

33   Lai CN, Gallick GE, Arlinghaus RB, Becker FF: Temperature-dependent transmembrane potential changes in cell infected with temperature sensitive Moloney sarcoma virus. J Cell Physiol 1984 (121):139-142

34 Binggeli R and Cameron IL: Cellular potentials of normal and cancerous fibroblasts and hepatocytes. Cancer Res 1980 (40):1830-1835

35 Koch KS and Leffert HL: Growth control of differentiated adult rat hepatocytes. In: Borek C and Williams GM (eds) Differentiation and Carcinogenesis in Liver Cell Cultures. Ann NY Acad Sci 1980 (349):111-127

36 Funkhouser WK, Pilch YH, Davies RJ: The electrophysiologic changes associated with premalignancy in colon carcinogenesis. Fed Proc 1986 (45):742

37 Davies RJ: Cell membrane depolarization in colon cancer. (manuscript in preparation)

38 Cameron IL, Smith NKR, Pool TB, Sparks RL: Intracellular concentration of sodium and other elements as related to mitogenesis and oncogenesis in vivo. Cancer Res 1980 (40):1493-1500

39 Zs-Nagy I, Lustyik G, Zs-Nagy V, Zarandi B, Bertoni-Freddari C: Intracellular Na+:K+ ratios in human cancer cells as revealed by energy dispersive X-ray microanalysis. J Cell Biol 1981 (90):769-777

40 Davies RJ, Weidema WF, Sandle GI, Palmer LI, Deschner EE, DeCosse JJ: Sodium transport in a mouse model of colonic cancer. Cancer Res 1987 (47):4646-4650

41 Furuya K, Enomoto K-I, Furuya S, Yamagishi S, Edwards C, Oka T: Single calcium-activated potassium channel in cultured mammary epithelial cells. Pflugers Arch 1989 (414):118-124

42 Enomoto K-I, Cossu MF, Edwards C, Oka T: Induction of distinct types of spontaneous electrical activities in mammary epithelial cells by epidermal growth factor and insulin. Proc Natl Acad Sci USA 1986 (83):4754-4758

43 Sellin JH: Intestinal electrolyte absorption and secretion. In: Sleisenger MH and Fordtran JS (eds) Gastrointestinal Diseases: Pathophysiology, Diagnosis, Management. WB Saunders, Philadelphia 1993 pp 954-976

44 Shennan DB: Mechanisms of mammary gland ion transport. Comp Biochem Physiol 1990 (97A):317-324

45 Shen SS, Hamamoto ST, Bern HA, Steinhardt RA: Alteration of sodium transport in mouse mammary epithelium associated with neoplastic transformation. Cancer Res 1978 (38):1356-1361

46 Marino AA, Morris DM, Schwalke MA, Iliev IG, Rogers S: Electrical-potential measurements in human breast cancer and benign lesions. Tumor Biol 1994 (in press)

47 Newcomb PA, Storer BE, Longnecker MP, Mittendorf R, Greenberg ER, Clapp RW, Burke KP, Willett WC, MacMahon B: Lactation and a reduced risk of premenopausal breast cancer. New Eng J Med 1994 (330):81-87

48 Greer FR, Tsang RC, Seary JE, Levin RS, Steichen JJ: Mineral homeostasis during lactation - relationship to serum 1,25-dihydroxyvitamin D, 25-hydroxyvitamin D, parathyroid hormone, and calcitonin. Am J Clin Nutr 1982 (36):431-437

49 Berga SE and Neville MC: Sodium and potassium distribution in the lactating mouse mammary gland in vivo. J Physiol 1985 (361):219-230

50 Kinura T: An electron microscopic study of the mechanism of milk secretion (translated). J Japan Obstet Gynecol Soc 1969 (21):301

51 Linzell JL and Peaker M: The permeability of mammary ducts. J Physiol 1971 (216):701-716

52 Linzell JL and Peaker M: Intracellular concentrations of sodium, potassium and chloride in the lactating mammary gland and their relation to the secretory mechanism. J Physiol 1971 (216):683-700

53 Blatchford DR and Peaker M: Effect of ionic composition of milk on transepithelial potential in the goat mammary gland. J Physiol 1988 (402):533-541

53 Blatchford DR and Peaker M: Effect of ionic composition of milk on transepithelial potential in the goat mammary gland. J Physiol 1988 (402):533-541

54 Wegman EA, Young JA, Cook DI: A 23-pS Ca2+-activated K+ channel in MCF-7 human breast carcinoma cells: an apparent correlation of channel incidence with the rate of cell proliferation. Pflugers Arch 1991 (417):562-570

55 Loo DD and Kaunitz JD: Ca2+ and cAMP activate K+ in the basolateral membrane of crypt cells isolated from rabbit distal colon. J Membrane Biol 1989 (110):19-28

56 Miller WR, Dixon JM, Scott WN, Forrest AP: Classification of human breast cysts according to electrolyte and androgen conjugate composition. Clin Oncol (R Coll Radiol) 1983 (9):227-232

57 Hess JC, Sedghinasab M, Moe RE, Pearce LA, Tapper D: Growth factor profiles in breast cyst fluid identify women with increased breast cancer risk. Am J Surg 1994 (167):523-530

58 Lewis SA, Eaton DC, Clausen C, Diamond JM: Nystatin as a probe for investigating the electrical properties of a tight epithelium. J Gen Physiol 1977 (70):427-440

59 Goller DA, Weidema WF, Davies RJ: Transmural electrical potential difference as an early marker in colon cancer. Arch Surg 1986 (121):345-350

60 Marino AA, Iliev IG, Schwalke MA, Gonzalez E, Marler KC, Flanagan CA: Association between cell membrane potential and breast cancer. Tum Biol 1994 (15):82-89

61 Davies RJ, Joseph R, Kaplan D, Juncosa RD, Pempinello C, Asbun H, Sedwitz MM: Epithelial impedance analysis in experimentally induced colon cancer. Biophys J 1987 (52):783-790

62 Morimoto T, Kinouchi Y, Iritani T, Kimura S, Konishi Y, Mitsuyama N: Measurement of the electrical bioimpedance of breast tumors. Eur Surg Res 1990 (22):86-92

63 Thurnherr N, Deschner EE, Stonehill EH, Lipkin M: Induction of adenocarcinomas of the colon in mice by weekly injections of 1,2-dimethylhydrazine. Cancer Res 1973 (33):940-945

64 Slaughter DP, Southwick HW, Smejkal W: "Field cancerization" in oral squamous epithelium: Clinical implications of multicentric origin. Cancer 1953 (6):963-968

65 Lippman SM, Lee JS, Lotan R: Biomarkers as intermediate endpoints in chemoprevention trials. JNCI 1990 (82):555-560

66 Muir R: The evolution of carcinoma of the mamma. J Pathol Bact 1941 (L11):155-172

67 Gallager HS and Martin JE: Early phases in the development of breast cancer. Cancer 1969 (24): 1170-1178

68 Schwartz GF, Patchesfsky AS, Feig SA, Shaber GS, Schwartz AB: Multicentricity of non-palpable breast cancer. Cancer 1980 (45):2913-2916

69 Ringberg A, Palmer B, Linell F, Rychterova V, Ljungberg O: Bilateral and multifocal breast carcinoma. A clinical and autopsy study with special emphasis on carcinoma in situ. Eur J Surg Onc 1991 (17):20-29

70 Anastassiades O, Iakovou E, Stavridou N, Gogas J, Karameris A: Multicentricity in breast cancer. A study of 366 cases. Am J Clin Pathol 1993 (99): 238-243

71 Alpers CE and Wellings SR: The prevalence of carcinoma in situ in normal and cancer associated breasts. Hum Pathol 1985 (16):796-807

72 Nielsen M, Christensen L, Andersen J: Contra-lateral cancerous breast lesions in women with clinical invasive breast cancer. Cancer 1986 (57):897-903

73 Simpson HW, Mutch F, Halberg F, Griffiths K, Wilson D: Bi-modal age-frequency distribution of epitheliosis in cancer mastectomies: relevance to preneoplasia. Cancer 1982 (50):2417-2422

74 Skolnick MH, Cannon-Albright LA, Goldgar DE et al: Inheritance of proliferative breast disease in breast cancer kindreds. Science 1990 (250):1715-1720

75 Rosen PP: Proliferative breast "disease": an unresolved diagnostic dilemma. Cancer 1993 (71):3798-3807

76 Wellings SR, Jensen HM, Marcum RC: An atlas of subgross pathology of the human breast with special reference to possible premalignant lesions. JNCI 1975 (55):521-535

77 Deschner EE: The Relationship of proliferative defects in colon cancer. In: Beahrs OH, Higgins GA, Weinstein JJ (eds) Colorectal Tumors. Lippincott Co, Philadelphia 1986 pp 81-86

78 Terpstra OT, Van Blankenstein M, Dees J, Eilers GAM: Abnormal pattern of cell proliferation in the entire colonic mucosa of patients with colon adenoma or cancer. Gastroenterology 1987 (92): 704-708

79 Lipkin M: Biomarkers of increased susceptibility to gastrointestinal cancer: New applications to studies of cancer prevention in human subjects. Cancer Res 1988 (48):235-245

80 Pretlow TP, Barrow BJ, Ashton WS: Aberrant crypts: putative preneoplastic foci in human colonic mucosa. Cancer Res 1991 (51):1564-1567

81 Flaxman BA and Lasfargues EY: Hormone-independent DNA synthesis by epithelial cells of adult human mammary gland in organ culture. Proc Soc Exp Biol Med 1973 (143):371-374

82 Allan DJ, Howell A, Roberts SA et al: Reduction in apoptosis relative to mitosis in histologically normal epithelium accompanies fibrocystic change and carcinoma of the premenopausal human breast. J Path 1992 (167):25-32

83 Van Bogaert LJ: The proliferative behavior of the human adult mammary epithelium. Acta Cytologica 1979 (23):252-257

84 Van Bogaert LJ: Effect of hormones on human mammary duct in vitro. Horm Metab Res 1978 (10): 337-340

85 Meyer JS and Connor RE: Cell proliferation in fibrocystic disease and post-menopausal breast ducts measured by thymidine labeling. Cancer 1982 (50):746-751

86 Ferguson DJP and Anderson TJ: Morphological evaluation of cell turnover in relation to the menstrual cycle in the "resting" human breast. Br J Cancer 1984 (44):177-181

87 Potten CS, Watson RJ, Williams GT, Tickle S, Roberts SA, Harris M, Howell A: The effect of age and menstrual cycle upon proliferative activity of the normal human breast. Br J Cancer 1988 (58): 163-170

88 Christov K, Chew KL, Ljung B-M, Waldman FM, Duarte LA, Goodson WH, Smith HS, Mayall BH: Proliferation of normal breast epithelial cells as shown by in vivo labeling with bromodeoxyuridine. Am J Pathol 1991 (138):1371-1377

89 Weiss BA, Ganepola GA, Freeman HP, Hsu Y-S, Faupel ML: Surface electrical potentials as a new modality in the diagnosis of breast lesions - A preliminary report. Breast Dis 1994 (7):91-98

90 Howell A, Anderson E, Laidlaw L, Schor A, Schor S, Potten C: Cyclical activity and "ageing" of the human breast: Clues to assessment of risk and strategies for prevention. In: Howell A (ed) Endocrine Therapy of Breast Cancer VI. ESO Monographs, Springer-Verlag, Berlin Heidelberg 1994 pp 27-46

91 Spring-Mills E and Elias JL: Cell surface differences in ducts from cancerous and non-cancerous breasts. Science 1975 (188):947-949

92 Courtney FP, Williams S, Mansell RE: Monoclonal antibody 323/A3, an indicator for the presence of breast carcinoma. Cancer Lett 1991 (57):115-119

93 Ceriani RL, Peterson JA, Blank EW: Variability in surface antigen expression of human breast epithelial cells cultured from normal breast, normal tissue peripheral to breast carcinomas and breast carcinomas. Cancer Res 1984 (44):3033-3039

94 Personal communication Dr W Dooley, September 1994

95 Jones JL, Critchley DR, Walker RA: Alterations of stromal protein and integrin expression in breast - a marker of premalignant change? J Pathol 1992 (167):399-406

96 Simpson HW, Griffiths K, McArdle C, Paulson AW, Hume P and Turkes A: The luteal heat cycle of the breast in health. Breast Cancer Res Treat 1993 (27):239-245

97 Simpson HW et al: Abstract presented at EUSOMA meeting, Venice, Italy, March 1994

98 Telang NT, Suto A, Wong GY: Induction by estrogen metabolite 16α-hydroxyestrone of genotoxic damage and aberrant proliferation in mouse mammary epithelial cells. JNCI 1992 (84):634-638

99 Osborne MP, Bradlow HL, Wong GY, Telang NT: Upregulation of estradiol C16α-hydroxylation in human breast tissue: a potential biomarker of breast cancer risk. JNCI 1993 (85):1917-1920

100 Weber G: Enzymology of cancer cells. New Engl J Med 1977 (297):486-493

101 McDermott EW, Barron ET, Smyth PP, O'Higgins NJ: Premorphological metabolic changes in human breast carcinogenesis. Br J Surg 1990 (77):1179-1182

102 Whittaker J, Walker RA, Varley JM: Differential expression of cellular oncogenes in benign and malignant human breast tissue. Int J Cancer 1986 (38):651-655

103 Jensen HM, Chen I, DeVault MR, Lewis AE: Angiogenesis induced by "normal" human breast tissue: A probable marker for precancer. Science 1982 (218):293-295

104 Schor AM, Schor SL, Ferguson JE, Rushton G, Howell A, Ferguson MWJ: Heterogeneity in the stromal components of the mammary gland: relevance to the development of breast disease. In: Mansel RE (ed) Benign Breast Disease. Parthenon Publishing Gp, Carnforth 1992 pp 115-129

105 Trosko JE, Chang CC, Madhukar BV: Chemical, oncogene and growth factor inhibition of gap junction intercellular communication: an integrative hypothesis of carcinogenesis. Pathobiology 1990 (58):265-278

106 Trosko JE and Chang CC: Gap junctional intercellular communication in neoplasia: implications for the cause and treatment of cancer. In: Herrera L (ed) Familial Adenomatous Polyposis. AR Liss, New York 1990

A History of Direct Current Measurement as it Relates to Tissue Proliferation

Mark L. Faupel

The University of Georgia, Boyd Graduate Studies, Research Center, Brooks Drive, Athens, GA 30602, U.S.A.

One of the central tenets of evolutionary biology is the conservation of morphological and behavioural characteristics in response to environmental demands. When challenged by the demands of a changing ecosystem, organisms typically adapt by successive modifications of existing biological machinery rather than the development of entirely new systems. A case in point involves the use of metabolic machinery which is responsible for the gradient maintained between a cell's internal and external environment. All differentiated cells actively maintain a selective ionic gradient across the cell membrane. When these cells are aggregated at the tissue level, a combination of epithelial cell-to-cell cohesion and polarity in the distribution of ionic pumps create and maintain appropriate internal salt and fluid levels within the tissue with respect to the external environment.

These tissue ionic gradients result from a functional asymmetry in the distribution of cell surface proteins. The vectorial transport of ions across epithelia generates significant transepithelial electrical potentials in a manner analogous to the charge separation maintained in inorganic batteries. In the epidermis of the adult frog, for example, substantial electrical potential differences (about 100 mV) are generated as Na enters the apical side of each cell and is extruded at the basal side, resulting in net Na transport in an apicobasal direction across the stratum granulosum [1]. This transport is mediated by Na-K ATPase at the basal aspect of the epithelia, and by amiloride-sensitive Na channels at the apical side of the epithelia. The function of this "pump" is to scavenge sodium from pond water, and can be thought of as an evolutionarily primitive adaptation to life in a low-sodium environment.

Data from Animals

If one examines higher functions, such as the neural crest formation in frog embryos [2], or synaptic function in adult organisms, the bioelectric phenomena are remarkably similar, involving the same ions and driving mechanisms. For example, synaptic activity of neurons involves saltatory depolarization along the axon to the terminal (presynaptic) bouton, where neurotransmitters are released. These neurotransmitters have an excitable effect on the post-synaptic membranes of neighbouring neurons which causes changes in membrane permeability to various ions. The ion species involved in neural transduction have been well studied and include calcium as the intracellular trigger of membrane permeability changes which alters sodium and potassium concentrations across the cell membrane. What is somewhat surprising at first glance is that the electrical activity of electrically inexcitable cells (such as epithelial cells) parallels very closely that of electrically excitable cells (e.g., neurons) [3]. As an epithelial cell divides, it depolarizes much as a neuron does when it fires. Calcium appears to be the intracellular trigger which alters membrane permeability to sodium and potassium [4]. Moreover, dividing cells and activated neurons feature the same degree of depolarization. Within an evolutionary context, it would seem that the relatively primitive ability of electrically inexcitable cells to manipulate internal and external salt and fluid balance has been conserved, but modified in electrically excitable cells to allow communication at a distance between cells.

DC versus AC Current Measurement

As applied to clinical medicine, two related technologies, electroencephalography and electrocardiology, capitalize on the aggregate of cellular action potentials generated by the brain and heart. In these technologies, periodicity of electrical activity, which corresponds to synchronous cellular activity, is the key diagnostic parameter. The informative signal is acquired using devices which measure alternating current. In contrast, electrically inexcitable cells do not demonstrate synchronous activity when they divide, so that measurement systems which can detect the lower amplitude, steady-state (or direct-current) activity associated with proliferation are necessary for diagnostic applications.

Preclinical and Clinical Data

This chapter briefly summarizes research findings which have demonstrated the relationship between steady-state electrical potentials and biological growth. The focus is on the history of dc current measurement with regard to organ systems. In the chapter by R.J. Davies in this volume the underlying cellular mechanisms that account for observed correlations between alterations in surface electrical potentials and tissue proliferation are discussed.

It has been nearly a century since Albert Matthews observed a correlation between biological growth and steady-state electrical fields. He measured consistent direct current voltage differentials of about 5 mV between the rapidly dividing polyp and the more quiescent stolon of hydroids [5]. However, E.J. Lund [6,7], who employed meticulous experimental techniques to document the presence of dc current in such species as *Obelia*, is generally given credit as the pioneer of steady-state current measurement [8].

In the 1930s and 40s, H.S. Burr, working at Yale University, carried out numerous experiments in which dc voltages were correlated with various growth phenomena, including neoplastic processes in rodents [9,10]. This line of research culminated in a human clinical study in which electropotential measurements were taken from the cervix of women with a variety of benign and malignant pelvic diseases [11]. Langman and Burr reported that their technique, which employed a measuring electrode on the cervix and a reference electrode on the ventral abdominal wall, could discriminate effectively between malignant and benign processes. Specifically, they claimed that negative polarity between the measuring and reference electrodes was diagnostic of cancer, while positive polarity between the measuring and reference electrodes was diagnostic of benign states. Burr attempted to account for these findings with an all encompassing theory, which was subsequently met by intense skepticism. Burr became somewhat of a tragic figure; a holistic-conceptual dinosaur in a world of increasing scientific reductionism. His failure to conceive of a mechanism to explain his empirical findings which was palatable to the scientific community led to apathy and eventual disinterest in his techniques.

It wasn't until the 1970s that interest in growth states and steady-state electrical fields re-emerged. The catalyst was the invention of the vibrating probe [12], a type of sensor system capable of accurately detecting small extracellular electrical currents in a variety of *in vivo* and *in vitro* systems. Among the reported findings using this technique are that both developing and mature epithelial cells drive a positive apical-basal current across themselves [13] and that these epithelial potentials are amplified during wound healing [14-16].

In 1982, Morris and Hirschowitz, inspired by Burr's work, reported results from a small study in which breast surface potentials were measured in patients scheduled for breast biopsy [17]. Their results were inconsistent, with some of the smaller cancers conforming to Burr's conception of negative polarity, but many of the cancers also produced breast potential measurements similar to those of benign lesions, that is, they were positive in polarity with respect to the reference electrode.

At about the same time, Nordenström [18] published a comprehensive volume which summarized his experience in measuring and manipulating internal dc fields. In this book, he describes evidence to support his concept of "biologically closed electric circuits", which he envisioned as an additional bioelectric circulatory system intimately tied to growth, development and healing. Nordenström's later work became focussed on treating cancer by the imposition of electrical current. His concepts and techniques have not yet been widely adopted in Western medical circles. To some extent, Nordenström's work parallels that of Burr: extensive but controversial empirical findings combined with theories that stretched the bounds of current medical thought which have resulted in a high degree of resistance to his views.

DC Measurements and Breast Cancer

Fuelled in part by an increased focus in applying biotechnology to *in vivo* diagnostic testing, the interest in studying surface electrical potentials for cancer diagnosis was rekindled. This interest was driven in part by the realization that early detection of certain cancers was important to reduce mortality and morbidity. In an attempt to diagnose breast cancer more accurately, a plethora of new presurgical tests have emerged to supplement mammographic screening and physical examination. These included thermography, ultrasound, magnetic resonance imaging, and various cellular sampling techniques, such as fine-needle aspiration cytology. With regard to breast cancer detection, the search is on for a low-cost non-invasive method which could reliably distinguish between malignant and non-malignant states.

In this regard, skin surface electropotential analysis appears to offer some of the attributes of a good screening or diagnostic test. The measurement is non-invasive, repeatable without risk, inexpensive, and does not require the infrastructural costs associated with many other procedures. Results are quantitative and do not require an expert to interpret them. Thus the focus of recent feasibility studies has been to identify the operating characteristics of this test and, in doing so, maximize its sensitivity and specificity.

One key realization regarding this operating system was that skin surface potentials are not precisely steady-state signals. Individual potentials can and do drift slowly over time, depending on internal and external physical states as well as on where the reference sensor is located. Thus any analysis of an absolute measure of depolarization must control for this inherent confound. The solution to this problem is to sample skin surface potentials from several sites of the organ system concurrently, so that voltages measured from one area of an organ system can be cross-referenced with voltages from other areas of the same organ system. DC drift over time ceases to be a problem because, within an organ system, voltages tend to drift in concert, a phenomenon which preserves the ability to assess relative levels of depolarization.

The importance of this effect was demonstrated in preliminary feasibility studies conducted in the US during the late 1980s [19]. Using computer-controlled, multiarray sensor systems, it was shown that relative differences in skin surface potential, both within the involved breast and between the two breasts, gave better diagnostic information than absolute levels of depolarization. In this study, electropotential measurements were taken from 104 suspicious lesions which were scheduled for biopsy. Of these 104 lesions, biopsy revealed that 36 were malignant while 68 were benign. Potential differentials both from within the involved breast and between the two breasts were significantly elevated in patients with breast cancer. A prospective diagnostic algorithm based on potential differentials produced a sensitivity of 97% and a specificity of 79%.

These encouraging initial results provided impetus to expand the clinical studies, both throughout the US and to Europe and Japan. Preclinical studies continue and should provide critical data to help understand the cellular mechanisms responsible for skin surface potential phenomena.

REFERENCES

1   Lindley BD: Fluxes across epithelia. Am Zool 1970 (10):355-364
2   Stump RF and Robinson KR: Ionic current in Xenopus embryos during neurulation and wound healing. In: Nucitelli R (ed) Ionic Currents in Development. Alan Liss Inc, New York 1986 pp 223-230
3   Putney JW: Excitement about calcium signaling in inexcitable cells. Science 1993 (262):676-678
4   Hoth M and Penner R: Identification of a calcium release-activated calcium current in electrically inexcitable cells. Nature 1992 (355):353-356
5   Matthews AP: Electrical polarity of the hydroids. Am J Physiol 1903 (8):294-299
6   Lund EJ: Experimental control of organic polarity by the electric current. II. The normal electric polarity of *Obelia*. A proof of its existence. J Exp Zool 1922 (36):477-494
7   Lund EJ: Electrical control of organic polarity in the egg of *Fucus*. Bot Gaz 1923 (76): 288-301
8   Lund EJ: Bioelectric Fields and Growth. University of Texas Press, Austin 1947
9   Nucitelli R: Introduction to ionic currents in development. In: Nucitelli R (ed) Ionic Currents in Development. Alan Liss Inc, New York 1986 pp 15-23
10 Burr HS, Strong LC, Smith GM: Bio-electric correlates of methyl-colanthrene-induced tumors in mice. Yale J Biol Med 1938 (10):539-545
11 Burr HS, Smith GM, Strong LC: Bioelectric properties of cancer-resistant and cancer-susceptible mice. Am J Canc 1938 (32):240-248
12 Langman L and Burr HS: A technique to aid in the detection of malignancy of the female genital tract. Am J Obstet Gyn 1949 (57):274-281
13 Jaffe LF and Nucitelli R: An ultrasensitive vibrating probe for measuring extracellular currents. J Cell Biol 1974 (63):614-628
14 Wiley LM and Nucitelli R: Detection of transcellular currents and effect of an imposed electric field on mouse blastomeres. In: Nucitelli R (ed) Ionic Currents in Development. Alan Liss Inc, New York 1986 pp 15-23
15 Barker AT, Jaffe LF, Vanable JW: The glabrous epidermis of cavies contains a powerful battery. Am J Physiol 1982 (242):R358-R366
16 Borgens RB: What is the role of naturally produced electric current in vertebrate regeneration and healng? Int Rev Cytol 1982 (76):245-298
17 Morris DM and Hirschowitz B: Electrical monitoring of breast carcinoma. J Bioelect 1982 (1):55-61
18 Nordenström B: Biologically Closed Electric Circuits. Nordic Medical Publications, Stockholm 1983
19 Weiss BA, Ganepola GAP, Freeman HP, Hsu Y-S, Faupel ML: Surface electrical potentials as a new modality in the diagnosis of breast lesions. A preliminary report. Breast Dis 1994 (7):91-98

Initial Preclinical Experiments with the Electropotential Differential Diagnosis of Mammary Cancer

"He saw what others had seen before him,
but it was only he who knew what it was he saw." *

David M. Long, Jr. [1,4], Mark L. Faupel [2], Yu-Sheng Hsu [3], Jorge A. Escobar [1],
Joseph P. Michel [1], Lawrence F. Mittag [1], Roxane M. Mitten [1], Brenda L. Witt [1],
William C. Herrick [4], and Robert E. Keefe [4]

1   Abel Laboratories, Inc., 2737 Via Orange Way, Suite 108, Spring Valley, CA 91978, U.S.A.
2   The University of Georgia, Boyd Graduate Studies, Research Center, Brooks Drive, Athens, GA 30602, U.S.A.
3   Georgia State University, Department of Math and Computer Science, 30 Prior Street, Atlanta, GA 30303, U.S.A.
4   Grossmont District Hospital-Sharp Healthcare, P.O. Box 158, La Mesa, CA 91944, U.S.A.

Direct current electropotential (EP) measurements were used in studying cancers in humans and experimental animals years ago, but this practice has not been accepted in the scientific community. Some studies, such as those of Burr and his colleagues [1,2], reported outstanding results with genital cancers in humans, but these studies were not pursued. Others, such as Morris and Hirschowitz [3], reported that the results of their clinical studies were erratic, with breast cancers smaller than 3 cm being electronegative while larger cancers showed either positive or negative EPs. Nordenström [4], the best known investigator in the field of cancer bioelectricity, has resorted to unconventional research techniques and publishing methods which are not subjected to peer review so that the entire body of his work is generally held in disrepute.

More recently the concept of using surface EP differentials based on measurements from an array of special sensors has been applied successfully in humans by Faupel et al. [5,6], resulting in remarkable accuracy in diagnosing breast cancer. This fundamentally different and highly specific principle has not been used in non-human species and has not been examined with the tools and methods available in the preclinical laboratory until this current work. In order to better understand and elucidate the basic scientific principles of surface EP measurements, we developed animal models which allowed assessment of a sensor system similar to the one applied successfully in humans.

This project was approached systematically by first experimenting with pairs of sensors, i.e., one measuring and one reference. We then tested systems of multiple sensors held rigidly in arrays, the position of which could be reproduced in a rodent over a series of days or weeks as mammary cancers developed either spontaneously or by implantation. Even though we were familiar with the clinical results, we were amazed at first to discover large EP differentials (difference between the highest and lowest EP) in regions of mammary tissue which was normal to palpation but revealed cancers 1 mm or less in diameter when dissected. We performed serial surface EP measurements in cancer-susceptible strains of C3H mice and found it was possible sometimes to predict areas where tumours would develop based on increased EP differentials. The results were similar to those reported from human experiments with breast cancer, although these preclinical tests were conducted independently of Faupel in a laboratory physically and administratively removed from his control but with his consultation.

---

* Quotation in part of the citation by W. Clarke Wescoe, M.D. who, as Chancellor and President of the Faculty of the University of Kansas, awarded the Distinguished Service Citation to the astronomer Clyde Tombaugh who discovered the long-sought planet Pluto when he was a teenager in New Mexico.

In order to replicate results, a high degree of skill and precise techniques were required. For example, it was difficult to adjust the pressure at the sensor-skin interface using rigid probes to support the tiny Ag/AgCl electrode discs. Spring-loaded probes to suspend the sensors were then added to our armamentarium. The lowest off-the-shelf 0.5 gram tension springs were associated with subtle changes from normal intercostal breathing to laboured diaphragmatic breathing after a few minutes when these probes were positioned over the sternum or chest wall. When customized 0.25 gram spring-loaded sensors were used, no changes in respiration were noted with up to 20 sensors placed over mammary tissue on the front, side and back of the chest. These sensors could be maintained in controlled positions for one hour on more than one occasion in the same mouse with no injury to the delicate skin or interference with breathing mechanics. Although it is not generally known, the mammary tissue of rodents has characteristic extensive distribution on the dorsal, ventral and lateral surfaces of the neck, chest and trunk [7]. Thus, a sensor system had to be devised to allow for these anatomical considerations.

We now felt prepared to undertake these pilot studies evaluating mammary cancer in rodents to determine the optimum size of the sensor. Also studied was the effect of different electro-conductive materials (ECM), such as gels and creams, used for completing the circuit between the skin and the Ag-AgCl electrodes. When we examined the preclinical and clinical bioelectrical data on mammary cancer, the patterns were reminiscent of earlier experiences from preclinical and clinical neuro-electrocardiographic observations in which chaos theory is applicable. For this reason, we submitted data to examination by non-linear neural network computer programmes. The preliminary results are intriguing and are therefore included. It may be possible to use neural networks in addition to other more conventional analytical methods to improve our diagnostic capabilities in order to approach the perfection in clinical diagnosis desired and expected by all potential breast cancer subjects.

At the same time as these experiments were being conducted in small animals, we were also pursuing applications in large animals where it is possible to apply and evaluate the same sensors and equipment used on humans. A retired dairy goat is the only animal with paired mammary glands approaching the dimensions of human breasts. Dairy goats are usually sociable and permissive to human contact. The purpose of this study was to develop a mammary cancer model in the goat by utilizing sustained-release Silastic® implants containing a known mammary carcinogen, 7,12-dimethyl-benzanthracene (DMBA), and to evaluate the efficacy of surface EP measurements in longitudinal studies utilizing this model. There are anatomical similarities between the human breast and udders, and the goat has a human-like immune system. The costs are also less than for other large animals, and there is a paucity of societal restraints compared with objections encountered when using primate and canine species.

Materials and Methods

Experiment A was performed on 12 eleven-week old female BALB/c mice, 18-20 g body weight, that were fed standard laboratory chow and water ad lib. Twenty-five thousand viable Experimental Mammary Tumour (EMT-6) adenocarcinoma cells and cell-free media were used. Viability was assessed by trypan blue exclusion method. Metofane anaesthesia was delivered by a gas delivery apparatus [8]. A platform was designed to hold the mice in a prone position so that an array plate of 3 rows of sensors could be positioned over the ventral surface of the mouse. Two of the rows covered each lateral side and one row covered the midline, so that the sensors covered the mouse from the neck to the symphysis pubis (see Fig. 1). In order to determine the optimal sensor in small rodents, the following sensors were used: 1) 4 mm diameter x 1 mm Ag-AgCl electrode discs, 2) 2 mm x 2.5 mm Ag-AgCl pellet and 3) 0.8 mm x 2.5 mm Ag-AgCl pellet. The within-subject design of Experiment A thus allowed comparison of sensor arrays consisting of either 4 mm, 2 mm, or 0.8 mm sensors. All sensors were purchased from IVM (Healdsburg, CA) and encased in probes containing springs of 0.25 ounce tension (Interconnect Devices, Kansas City, KS) which were then attached to the array plate. Synapse® electro-conductive cream (Med-Tek Corp., Northbrook, IL) was placed on the sensor before applying the sensor to the skin surface. A highly sensitive 32-channel volt meter with 1 MΩ impedance and a filter excluding all signals above 0.25 HZ (Biofield Meter, Biofield Corp., Roswell, GA) was used to take the EP readings. Mice were placed in a bell jar containing a Metofane® soaked gauze to induce anaesthesia. Their ventral surfaces were shaved with clippers as closely as possible. They were then placed on a platform with gas inhalation apparatus in place thus maintaining the anaesthesia level adequate to avoid motion. The sensor array was then positioned by lowering the plate holding the sensors and by screwing the sensor probes individually in an appropriate direction, clockwise or counterclockwise, for fine adjustment. The skin surface EP measurements were taken over approximately 3 minutes (a total of 50 times for each sensor), and the average EP was calculated and displayed in millivolts (mV) on the monitor screen of the built-in lap-top computer. In addition to the average EP, the variability (difference between the highest and the lowest of the 50 EP readings) and the reliability (percentage of the 50 EP readings used in calculating the average) were also displayed. EP measurements were obtained 3 times a week for 2 weeks to establish baseline EPs for each location on the body surface. Thereafter 6 mice were selected at random to receive an injection of 25,000 viable EMT-6 cells into the mammary gland, while the other 6 mice received cell-free media injections. Location of injection site was on the ventral surface of 1 of 4 mammary chain regions as follows: right or left pectoral or right or left inguinal. The data collector was blinded as to location of the injection. Data collections were made 3 times per week until tumours became approximately 1 cm in diameter after which the mice were sacrificed.

Mammary tissue was collected for histological examination by first covering the mouse with chemical depilator after euthanasia by CO_2 inhalation, washing the depilator off, and then removing skin containing mammary tissue and placing the tissue in a cassette which was placed in 10% buffered formalin. Fixed tissue sections embedded in paraffin were stained with hematoxylin and eosin and examined microscopically by a pathologist.

For each sensor examined, the best test was chosen by following the criteria of using sensor EP measurements having a variability or range less than 10 mV and a reliability greater than

Fig. 1. This drawing shows a BALB/c mouse and the current apparatus for measuring surface EPs in a mouse. The equipment contains a built-in thermal unit in the platform to maintain body temperature, and a 32-sensor array of 0.8 mm diameter Ag/AgCl sensors mounted on probes loaded with 0.25 ounce tension springs to avoid excessive pressure on the skin and not affect the breathing pattern of the mouse. The tips of the sensors are coated with Synapse® electroconductive cream. This apparatus permits simultaneous measurement of surface EP maps on ventral, dorsal and lateral aspects of a mouse and can be expanded to contain as many as 64 sensors. Preparation and performance of an individual study on a mouse can be accomplished in approximately 30-40 minutes.

Experiment A was performed using an array of 20 sensors on the ventral and lateral body surface with the mice supine. Experiment B was performed using a 20-sensor array positioned on the dorsal and lateral sides with the mice prone. The apparatus is designed to permit rapid positioning of the arrays and to allow for fine adjustment of the contact of each sensor on the surface of the skin. Continuous inhalation anaesthesia is accomplished with the head of the subject in the nose-cone delivering a volatile anaesthetic agent.

0.98; that is, a minimum of 98% of the voltage measurements were required for an averaged reading to be used. Any readings not within these criteria were considered unreliable and unstable due to failure of the sensor to equilibrate to the skin. If more than one test was made, the test with the lowest variability was chosen. Differentials were calculated by taking the difference between the highest and the lowest EP reading of a particular region. Differentials of the following regions were calculated: the pectoral area, the inguinal area, the right lateral and the left lateral areas, the midline region, the overall region, the overall region ex-

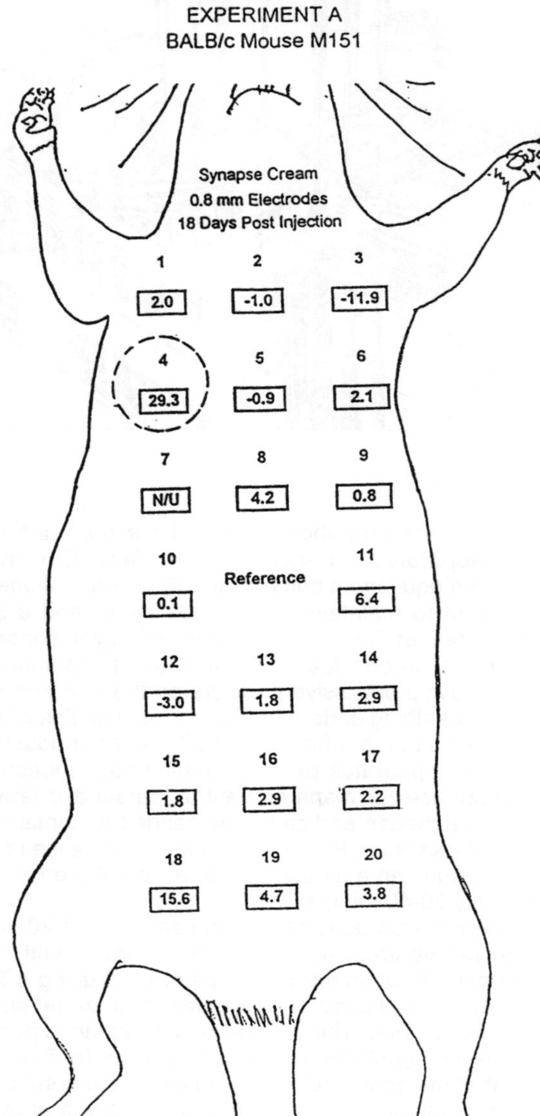

Fig. 2. This drawing illustrates a typical "voltage map" on a mouse in Experiment A. The sensor positions are labelled 1 through 20, and the EP at each location is given in millivolts. This reading was obtained at the time that the mouse had a mammary cancer located at the site of sensor 4 as indicated by the circular outline of the tumour. The term NU indicates this EP was not used due to unacceptable measurement variability. There was a 41.2 mV differential between sensors 3 and 4, a finding consistent with and diagnostic of the presence of a malignant tumour.

cluding the midline, and the average of the corresponding lateral sensors. For each region the pre- and the post-injection mean and standard deviation were calculated for the first sensor tested, the second sensor tested, and the third sensor tested, all 0.8 mm sensor readings, all 2 mm sensor readings, and all 4 mm sensor readings for each mouse.

In Experiment B, 27 BALB/c mice (Harlan Sprague Dawley, Mira Mesa, CA) were used in the second experiment testing 2 different types of electroconductive gels (Parker Laboratories, Inc., Orange, NJ). The 0.8 mm x 2.5 mm Ag-AgCl sensors encased in a probe with a 0.25 ounce tension spring were used. At the beginning of each day before testing any subjects, an EP recording was taken with the sensor array immersed in 5% saline solution. The sensors were depolarized by applying 2 mAmps current, 12 Volts AC, for 2 minutes with all sensors in the array grounded in 5% saline solution. After depolarizing sensors, and with the sensors still immersed in the saline solution, a second recording was taken. Thereafter, application of current to depolarize sensors was performed between each test subject but no data were recorded. After all test subjects were studied for the day, a Biofield Meter reading of the sensors immersed in 5% saline was taken.

Mice were placed on the rodent table platform fitted with a heating unit and anaesthetized with Isoflurane® using a gas anaesthesia delivery system. Induction time, time off anaesthesia and recovery time were recorded. Rectal temperature was recorded using a thermister probe and body temperature was controlled throughout the test period. The hair was removed from the areas of interest as needed using a clipper.

Signa spray (Parker Laboratories Inc., Orange, NJ) was applied as a skin prep by spraying it directly on the mouse, and after 30 seconds the skin was dried with a dry gauze pad. The ECM was applied to the sensors, and the sensors were applied to the areas of mammary tissue on the dorsal and lateral surfaces as shown in Figure 2. Once all sensors were in contact with the skin, a series of EP measurements were obtained. If at any time during the readings a sensor measurement did not meet criteria of 0.98 reliability with a variability of less than 10 mV, adjustments such as regelling or repositioning were made before continuing. All mice were tested twice per week with each ECM on alternating days. Twenty consecutive Biofield Meter tests were performed (approximately 1 hour) for the first day and a half. After analyzing these initial results, it was decided to reduce the study time to 7 tests or no more than 30 minutes. This step also reduced the stress related to prolonged anaesthesia. Six baseline studies were performed for each ECM on all mice. After the baseline studies were collected, 25,000 viable EMT-6 cells were injected subcutaneously into either the right or left pectoral mammary glands in the dorsal region. Data collectors were blinded as to the injection site. Surface electropotential readings were taken until the tumours reached a diameter of approximately 5 mm, and then the mice were sacrificed by inhalation of CO_2. The mammary gland and lung tissue were then prepared for histological examination.

For each sensor tested, the best test was chosen by following the criteria of using sensor measurements having a variability less than 10 mV and a reliability greater than 0.98. Any readings not within these criteria were considered unreliable and unstable due to failure of the sensor to equilibrate to the skin. If more than one test was made, the test with the lowest variability was chosen. For each ECM the mean of differentials, standard deviation and standard error of the mean were determined for each region over time for baseline and post-injection readings. The overall peak differential as well as the pectoral and inguinal region differentials were used to determine significance between both ECMs using the Max-T test to compare any changes in the t-distributions between pre-injection and post-injection differentials. The dimensions of the tumours were measured (length and width) with calipers from the time the tumours first became palpable. Tumours were graded following microscopic examination by a pathologist.

The Backpropagation configuration of Neuroshell 2 (Ward Systems Group, Inc., Frederick, MD) was used in the applications of neural networks. The inputs selected were the average raw EP values from the 20 sensors, the pectoral, inguinal, and overall differentials, the overall differential without the midline, the left, right and midline differentials, and the lateral differential values. The network was trained to recognize and distinguish patterns of EPs and EP differentials from data collected before the tumour cells were injected and to compare these with data collected after injection of tumour cells. Neural network training required two weeks using the inputs from tests 3 and 4. Once trained, the network was given the inputs from tests 5, 6 and 7 which it had not seen before. A score of 0 was assigned to the pre-injection tests and a score of 1 was assigned to the post-injection tests so that the closer the score to 0 or 1, the better the network was able to distinguish between the EP measurements taken in the pre-injection phase versus the post-injection phase of tumour growth.

Experiment C was conducted on retired dairy goats in order to apply the same sensors used in human studies and to develop a model that would permit longitudinal studies throughout the life history of a mammary cancer. In this study each of two goats had Silastic® implants (approximately 250 µl Silastic with a trace of plasticizer) containing 50 mg DMBA injected in a semi-liquid state into one udder at a 1.0 cm depth which was then allowed to solidify; the opposite udder was implanted with Silastic alone (Control, no DMBA). The data collectors were blinded as to which udder received the DMBA.

EP readings were obtained from each udder simultaneously using an 11 sensor array on each udder with the pattern diagrammed below (electrodes 1-11 and 17-27). Sensors were placed on the lower abdominal wall over the outer sciatic lymph nodes bilaterally which have about the same relationship to the udders as the axillary lymph nodes have for the human breast. One reference sensor was placed on the left lateral thigh making a total of 24 sensors plus one reference sensor.

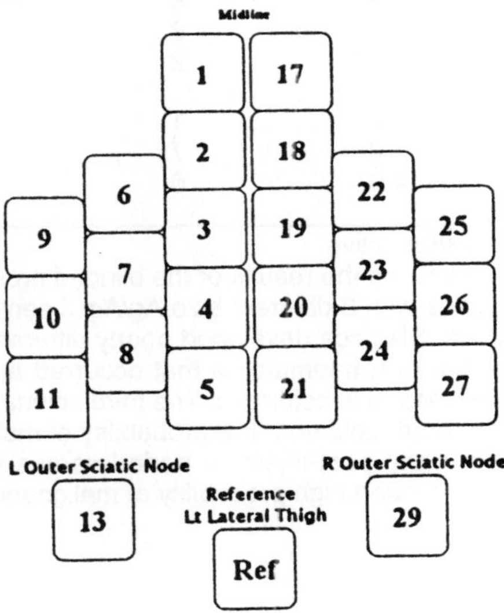

Goat Sensor Array

Table 1. Results of blinded prospective analysis

Mouse No.	Electrode size (mm)	N	Probability of malignancy	Pathology results	Injection site
M144	0.8	8	high	poorly	right pectoral
	2.0	6	high	differentiated	
	4.0	7	high	carcinoma	
M146	0.8	8	high	poorly	left pectoral
	2.0	6	high	differentiated	
	4.0	7	high	carcinoma	
M151	0.8	5	moderate	poorly	right pectoral
	2.0	4	moderate	differentiated	
	4.0	2	low	carcinoma	
M157	0.8	8	high	poorly	left pectoral
	2.0	3	moderate	differentiated	
	4.0	7	high	carcinoma	
M159	0.8	7	high	poorly	left pectoral
	2.0	7	high	differentiated	
	4.0	7	high	carcinoma	
M150*	0.8	8	high	normal	right pectoral
	2.0	4	moderate	mammary	
	4.0	7	high	tissue	
M153	0.8	0	low	normal	left pectoral
	2.0	0	low	mammary	
	4.0	0	low	tissue	
M155	0.8	1	low	normal	left pectoral
	2.0	5	moderate	mammary	
	4.0	6	high	tissue	
M156*	0.8	7	high	inflammation,	left pectoral
	2.0	7	high	congestion and	
	4.0	7	high	some necrosis	
				mammary tissue	
M158	0.8	5	moderate	normal	left pectoral
	2.0	2	low	mammary	
	4.0	2	low	tissue	
M161	0.8	1	low	normal	right pectoral
	2.0	7	high	mammary	
	4.0	6	high	tissue	

* means false positive

This table contains the results of the blinded prospective randomized analysis of the EP differentials measured in the 11 surviving mice with 3 different-size Ag/AgCl sensors in order to determine the probability of an abnormality in the mammary tissues. Six mice developed poorly differentiated mammary carcinomas, and one mouse had an inflammatory reaction related to a haematoma that occurred at the site of injection of cell-free media. The diameter of the sensor is given in millimeters in column 2. The third column labelled N gives the number of increases in differential for each sensor size. In the fourth column a low probability of malignancy was scored when 0-2 was the number of increases in differential means in the post-injection period versus the pre-injection period. 3-5 increases was rated moderate, and 6-8 increases was rated high probability of malignancy.

Table 2. Statistical comparison of sensor size

Sensor size	Sensitivity	Specificity
0.8 mm	100%	60%
2.0 mm	100%	40%
4.0 mm	83%	40%

This table gives the statistically determined sensitivity and specificity of the presence of abnormalities in the mammary tissue as determined by EP differentials measured in mice receiving injections of either 25,000 viable EMT-6 cancer cells or cell-free media in the case of sham-treated controls in Experiment A. In this pilot study the 0.8 mm diameter Ag/AgCl sensors had the best scores for predicting the abnormalities in the tissues.

EP readings were obtained twice a week in the control period. After the DMBA capsules were implanted the EP readings were obtained once a week. Postmortem examinations were made on all goats sacrificed with lethal doses of Beuthanasia®, and histological analysis was performed on selected tissue from each mammary gland.

Results

Experiment A

Five mice developed poorly differentiated mammary carcinoma at the site of injection of the EMT-6 tumour cells. The sixth tumour-injected mouse died shortly after injection of tumour cells. In 5 of the mice that received cell-free media injections, there were no abnormalities of the mammary tissue. One of the 6 control mice developed a large haematoma immediately after injection of cell-free media. The haematoma resolved rapidly, but at microscopic postmortem examination, a moderately severe inflammatory reaction with tissue congestion and necrosis was seen at the injection site. Thus, only 5 of the sham-injected mice had normal mammary tissue.

Table 1 summarizes the results of a blinded prospective analysis of the 8 regions studied in the 11 surviving mice. The third column gives the number of times in which the differential of a region increased in the post-injection period. The fourth column is based on the number of increases of the post-injection versus the pre-injection differential means. The fifth column

gives histopathological results. Statistical examination of these results (Table 2) revealed that surface EP evaluation was capable of distinguishing normal from abnormal (cancer or inflammation) mammary tissue with a sensitivity of 100% and a specificity of 60% using 0.8 mm electrodes, while with the larger 4 mm sensor the results were not as good. In Figure 2 a typical voltage map of a tumour bearing mouse is illustrated showing a 41.2 mV differential between sensors 3 and 4 using the 0.8 mm sensor with a tumour located in the vicinity of sensor 4.

Experiment B

These experiments compared the effectiveness of two different ECMs for detecting malignant tumours. Based on a visual examination of data obtained measuring EPs over 1 hour, it was determined that the differentials increased through tests 1 to 7 with a leveling off of the rate of change in tests 5, 6 and 7 (Fig. 3). Thereafter the differentials tended to decrease in an erratic fashion, and were not used. In order to reduce exposure time to anaesthetics, the period of testing was reduced to include tests 5, 6 and 7, which typically occurred within 30 minutes. Complete information on pre- and post-injection EP differentials on both ECMs was obtained on 18 animals that survived the full experiment. Nine animals expired in the early stages of the experiment while we were developing skills in delivering anaesthesia, and their results are not included.

The peak differentials for each region were determined for each day before and after injection of tumour cells, and the mean, standard deviation and standard error were calculated for each region and ECM. Table 3 contains the means, SEs and Ns for this experiment. These data are expressed graphically in Figure 3 for the overall region. The overall peak differential (highest peak from tests 5, 6 and 7 as well as pectoral and inguinal differentials, average of tests 5, 6 and 7 combined) were used to find the significance between both ECMs using the Max-t test, comparing the pre-injection results with the post-injection results. This analysis is given in Table 4 which shows there was a highly significant difference between the differentials of the baseline and those of the tumour phase in the overall and pectoral regions but

Table 3. Summary of regional differential calculations

Pre-implant of mammary tumour

GEL "S"	Day 1	7	8	12	Mean			
Overall mean ± SE (N)	**35.8** 1.4 (18)	**38.6** 2.4 (18)	**35.2** 1.3 (18)	**37.1** 1.9 (18)	**36.7** 0.9 (72)			
Pectoral mean ± SE (N)	**33** 0.9 (54)	**31.6** 1.3 (1.3)	**29.7** 0.9 (54)	**33.2** 1.2 (54)	**31.9** 0.5 (215)			
Inguinal mean ± SE (N)	**26.5** 1 (54)	**27.1** 1 (53)	**25.6** 1.3 (54)	**27** 1.1 (54)	**26.6** 0.5 (215)			**TUMOUR IMPLANT**

Post-implant of mammary tumour

	Day 15	19	22	26	28	Mean	32	34
Overall mean ± SE (N)	**39.9** 2.2 (18)	**39.5** 3.2 (17)	**40.1** 2.2 (18)	**48.9** 2.6 (18)	**34.6** 1.9 (18)	**40.6** 1.1 (89)	**39.7** 4 (4)	**39.9** 4.2 (4)
Pectoral mean ± SE (N)	**34.2** 1.4 (54)	**32.9** 1.7 (49)	**35.1** 1.3 (53)	**41.9** 1.4 (53)	**31.2** 1.1 (54)	**35.1** 0.6 (263)	**36.2** 3.1 (11)	**35.9** 2.2 (12)
Inguinal mean ± SE (N)	**27.7** 1.3 (54)	**28.1** 1.6 (49)	**29.8** 1 (53)	**34.8** 1.3 (53)	**26.6** 1 (54)	**29.3** 0.6 (263)	**26** 1.7 (11)	**27.2** 1.2 (12)

Pre-implant of mammary tumour

GEL "P"	Day 2	8	9	13	Mean			
Overall mean ± SE (N)	**51.4** 3.8 (18)	**50.5** 2.4 (18)	**53.6** 2.5 (18)	**57** 4.4 (18)	**53.1** 1.6 (68)			
Pectoral mean ± SE (N)	**43.6** 2 (53)	**42.9** 1.5 (5.4)	**44.7** 1.6 (53)	**48.3** 2.9 (53)	**44.8** 0.9 (213)			
Inguinal mean ± SE (N)	**40.7** 1.2 (53)	**39** 1.2 (54)	**39.7** 1.6 (53)	**45.6** 1.7 (53)	**41.2** 0.7 (213)			**TUMOUR IMPLANT**

Post-implant of mammary tumour

	Day 16	20	23	27	29	Mean	33	34
Overall mean ± SE (N)	**56.1** 7.2 (18)	**52.8** 2.8 (18)	**58.9** 4 (18)	**56.1** 2.8 (18)	**59.4** 4.7 (18)	**56.7** 2 (90)	**67.8** 13 (4)	**54.7** 6 (4)
Pectoral mean ± SE (N)	**46.7** 4 (54)	**44.5** 1.6 (53)	**49.4** 1.9 (54)	**46.6** 2 (54)	**50.6** 2.7 (54)	**47.6** 1.1 (269)	**53.5** 3.6 (12)	**51.3** 2.8 (12)
Inguinal mean ± SE (N)	**37** 1.3 (54)	**37.7** 1.4 (53)	**42** 1.6 (54)	**41.8** 1.5 (54)	**44.4** 1.7 (54)	**40.6** 0.7 (269)	**45** 5.4 (12)	**32.5** 2.5 (12)

The calculations of the regional differentials from Experiment B are presented including the standard error of the mean and the number of studies used for each calculation before and after implant of the mammary cancer. The timing for each study is given in days after start of data collection.

Table 4. Max-t test results

Region of animal	Gel "S"	Gel "P"
Overall	P = 0.0000014	P = 0.0558458
Pectoral	P = 0.0035406	P = 0.1709365
Inguinal	P = 0.0916696	P = 0.4534083

The overall peak differentials as well as the pectoral and inguinal regional differentials from Experiment B were used to find the significance between both ECMs using the Max-t test which compares any changes in the t-distribution between pre-injection and post-injection differentials

Fig. 3. This bar graph illustrates the means with SE bars of the peak EP differentials of each of 7 consecutive EP tests over a period of approximately 30 minutes for each experimental day in 18 mice from Experiment B during the baseline period compared with the post-injection period of mammary tumour growth. The magnitude of the differentials was greater with Gel "P" than with Gel "S"; however, the differences between the baseline and the post-injection periods were significant only for the gel "S" in the second through seventh tests. Note that the differentials increased less in tests 5, 6 and 7 than in tests 1 through 4.

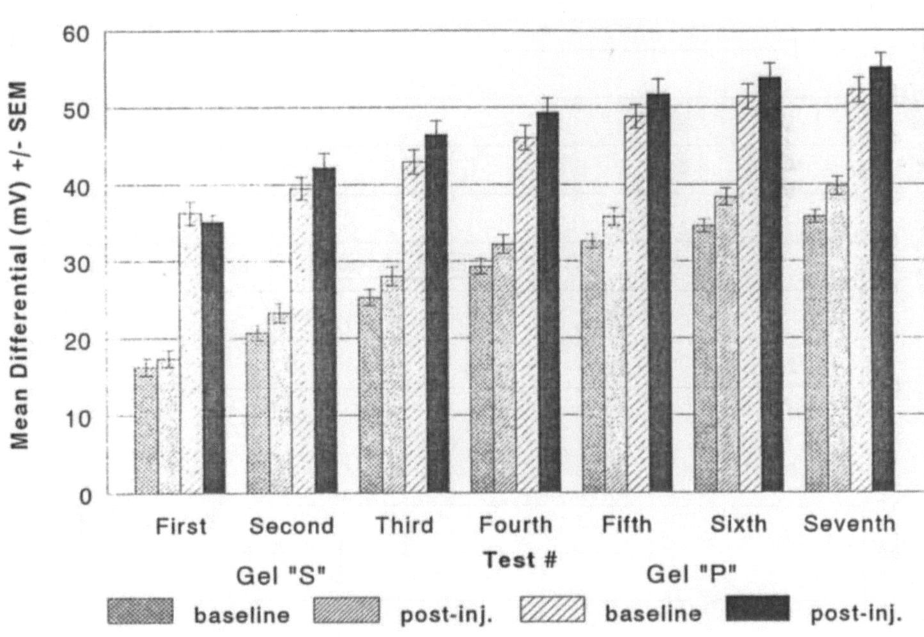

Fig. 4. This line graph illustrates the increase in the tumour area from the 22nd to the 53th day of Experiment B in which the EMT-6 tumour was implanted in the mammary gland in the right or left pectoral region on the 14th day. There were 18 mice included until the 29th day, at which time 14 mice with large tumours were euthanized, and the 4 mice with smaller tumours were euthanized on the 35th day.

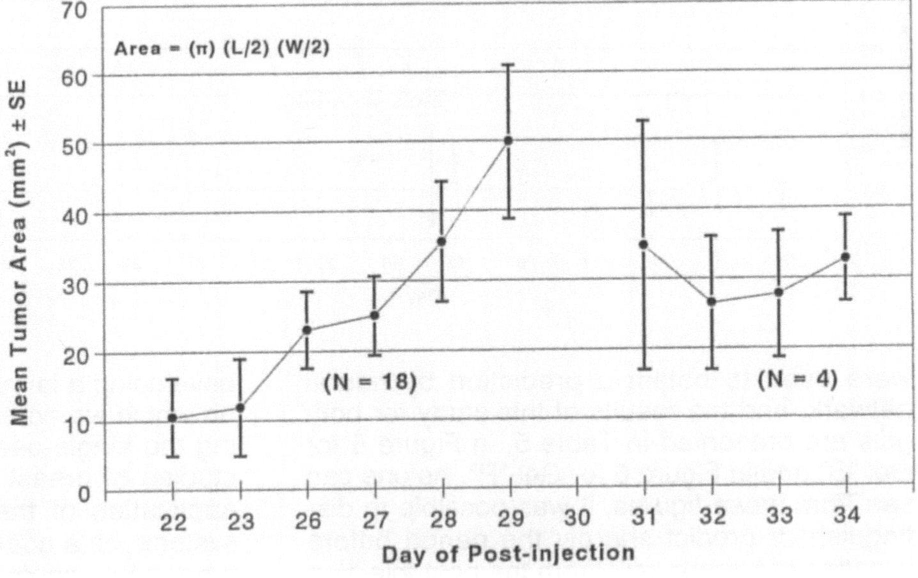

not in the inguinal region when the Gel "S" was used. When the Gel "P" was used, the differences in differentials for baseline versus tumour phases approached significance in the overall region but not in the pectoral or inguinal regions.

The changes in the surface area of the tumours are presented graphically in Figure 4 which demonstrates a progression of tumour size from the 22nd day to the 29th day, at which time 14 mice with large tumours were sacrificed. The measurements on the 4 mice with smaller tumours were continued for 6 more days.

The analysis of these data by the NeuroShell 2 neural network computer programme is a new undertaking and is ongoing, but preliminary results are interesting and promising. The raw data from EP measurements and differentials

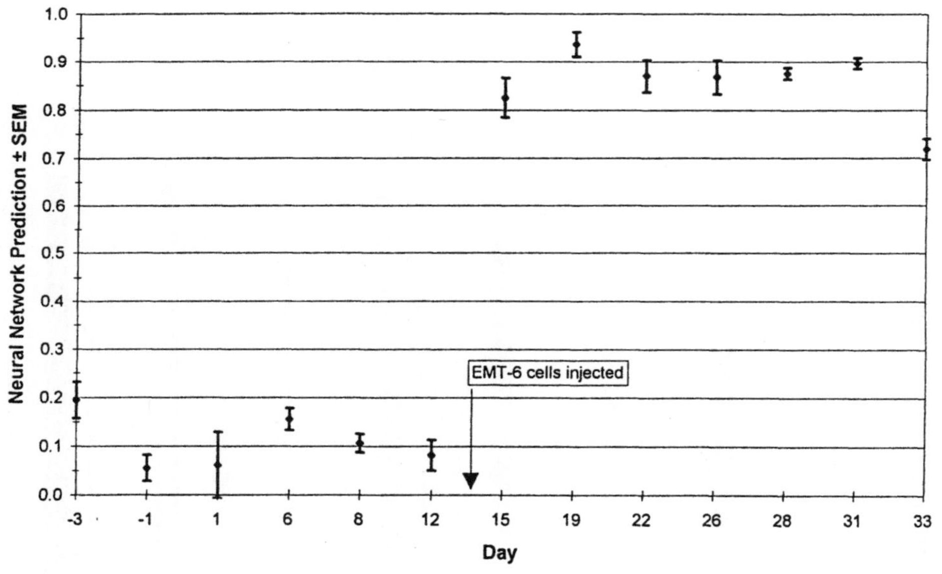

Fig. 5. This graph demonstrates the results of the neural network analysis of the EP data from Gel "S" of Experiment B. Note the abrupt increase in the score following the injection of tumour cells on day 14. (0 = baseline reading, 1 = post-injection reading)

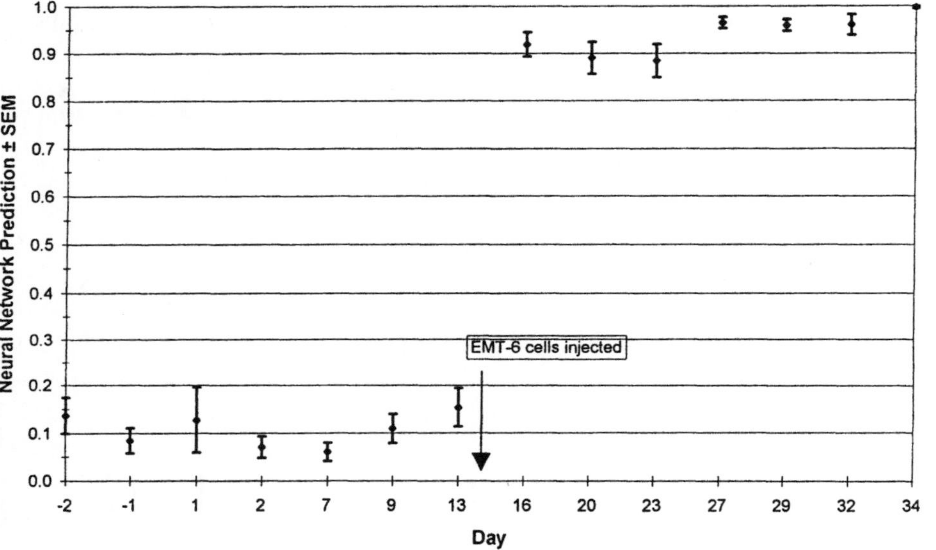

Fig. 6. This graph shows the results of the neural network analysis of the EP data from Gel "P" of Experiment B. Note the abrupt increase in the score after the tumour cells were injected on day 14 and the perfect score of 1 which was achieved with the data from the 34th day. (0 = baseline reading, 1 = post-injection reading)

were used to obtain a prediction by neural network, and the results of this study for both gels are presented in Table 5, in Figure 5 for Gel "S" and in Figure 6 for Gel "P". As one can see from these figures, it was possible to distinguish or predict sharply the period before injection of tumour cells from the post-injection period using this system of non-linear analysis. What is not known is the pattern or patterns recognized by this complex and novel technique in making such a precise definition of the time at which tumour cells were injected.

Experiment C

This goat model is still in a preliminary state; however, we were able to achieve the goal of developing a large animal experimental model on which we could conduct investigations using the single-use sensors applied in human studies of breast cancer. An example of the application of the human-type sensors and systems on a goat is seen in Figure 7. The retired dairy goats proved to be cooperative subjects with paired udders about the size of the average human breast.

An example of the differentials measured in a goat with DMBA and placebo implants is demonstrated in Figure 8. The carcinogen containing implant was in the left udder, and the right udder contained the placebo. As can be seen in these examples, the differentials increased for the first 90 days in both udders with either implant. The differentials measured

Table 5. Average neural network prediction over time

Gel "S"

	-3	-1	1	6	8	12	15	19	22	26	28	31	33
Average	0.19	0.05	0.06	0.16	0.11	0.08	0.82	0.94	0.87	0.87	0.88	0.90	0.72
SE	0.05	0.03	0.03	0.04	0.03	0.02	0.04	0.01	0.04	0.03	0.03	0.05	0.13
No. of readings	49	18	34	53	51	52	54	48	52	49	52	11	12

Gel "P"

	-2	-1	1	2	7	9	13	16	20	23	27	29	32	34
Average	0.14	0.08	0.13	0.07	0.06	0.11	0.15	0.92	0.89	0.88	0.96	0.96	0.96	1.00
SE	0.04	0.03	0.07	0.02	0.02	0.03	0.04	0.03	0.03	0.04	0.01	0.01	0.02	0.00
No. of readings	34	18	16	29	50	50	51	51	49	54	52	53	11	12

0 = baseline readings (pre-injection), 1 = post-injection readings. N = 18 mice
This chart represents the scores of the neural network predictions over time involving the 33 days of Experiment B. Note the low scores of 0.05 to 0.19 for days -3 to 13, which was before the tumour cells were injected on day 14. In the post-injection phase the scores of the neural network ranged from 0.72 to 1.0, which indicates a high confidence for predicting the differences in the input of EPs and EP differentials between the pre and post-injection phases.

after the initial burst decreased to about the same values seen in the control period. In the final measurement, the differential increased to 38 mV in the symptomatic left udder with a 22 mV differential in the control right udder. The difference between udders was 16 mV which may be indicative of an abnormality in the left udder.

Examination of the udders after implantation revealed that there was a palpable mass in both udders for the first 4 to 6 weeks. Thereafter the mass in the placebo-treated udder could not be felt, and that in the carcinogen treated udder increased. At postmortem examination the gross findings revealed that the implants were in connective tissue and fat superficial to the tough capsule surrounding the mammary glands. There was a large collection of rock-hard fibrous tissue and inflammatory cells associated with the carcinogenic implant, but the placebo containing udder showed only a mild fibrous reaction similar to that seen in implants of non-irritating polymers such as Silastic.

We did not achieve the goal of producing a mammary cancer in goats. The differentials measured increased over the first 90 days after implanting the capsule with DMBA or placebo, and thereafter the differentials decreased. After 460 days, the differentials in the DMBA treated mammary gland were consistently higher than in the control udder. Although cancer did not develop in this goat model, cancer did develop in 5 out of 5 rats after 6 to 11 months using DMBA-silastic implants in the mammary glands in preparatory pilot studies (unpublished data). The goat model was successful in the ability to obtain useful EPs from the skin surface over

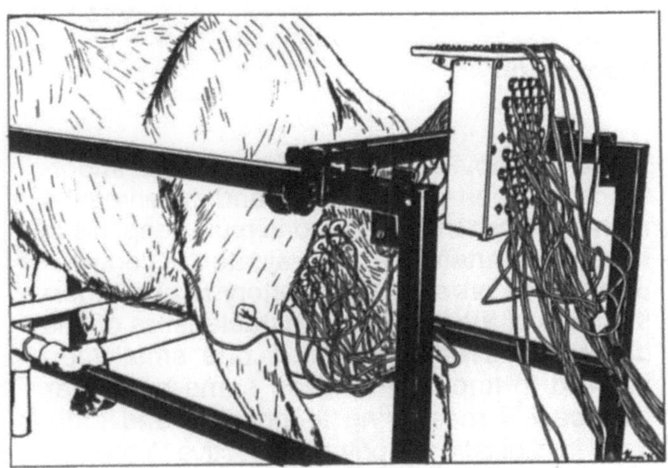

Fig. 7. This drawing shows a dairy goat in a milking stanchion without any sedation. Illustrated is an array of 22 single-use sensors over the mammary glands (udders), two sensors over each principal lymphatic structure draining the udder (outer sciatic node in the lower abdominal wall) and one sensor in a reference position (lateral thigh). The sensors are connected to standard Biofield lead wires through an intermediate device designed for the goat laboratory.

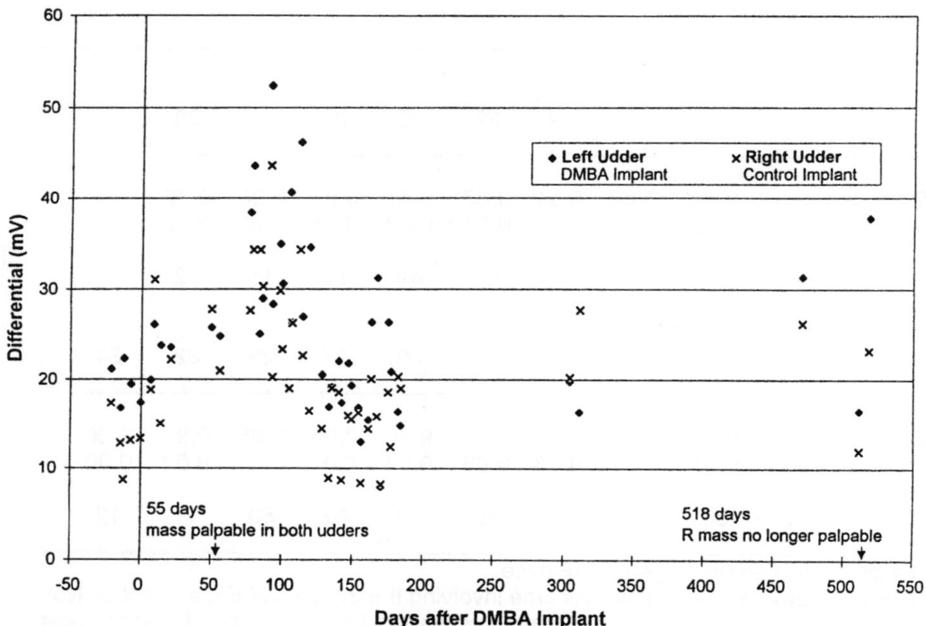

Fig. 8. This is a graphic presentation of EP differentials obtained from the retired dairy goat seen in the drawing in Figure 7. At time 0 she received silastic implants in both udders with the left udder implant containing a carcinogenic dose of DMBA. Note that there was a large increase in the differentials bilaterally in the first 90 days, following which the differentials decreased. The differentials were higher in the left udder during the last 48 days of the experiment, reaching a maximum of 14.5 mV.

mammary glands and to observe electrical changes consistent with an underlying pathological process. From this point of view, the goat studies were of some value, and the model may prove to be effective for testing new sensors in long-term studies. The model may also be useful in developing systems to image or localize lesions with electrical modalities alone.

Discussion

There are obvious differences of great importance between breast cancer and experimental mammary carcinoma, and one must keep this in mind when attempting to evaluate the contributions of the preclinical laboratory to our understanding of the biological characteristics of this disease. Nevertheless, it is one small step forward in knowing that the same or similar surface EP measuring techniques and tools used in clinical oncology are effective in preclinical oncology. Breast cancer is an emotionally-charged condition so there is little opportunity to purposely study the electrical life history of this malady in a controlled setting. Thus, one works with many snapshots taken from scores of patients with highly individualized breast pathology in a continuum of stages that can

never be completely defined and reproduced. The experimental laboratory results reported herein suggest there may be a tool that allows us to control and reproduce mammary cancer and correlate the surface electrical characteristics with the electrical potentials generated by underlying disease states. Furthermore, these tools may be adapted to examine the distortions of surface EP maps associated with deeply placed cancers such as ovarian and uterine cancer as reported by Langman and Burr [2]. Finally, it may be possible to use non-invasive, inexpensive EP measurements to accurately and reliably diagnose or predict physiological events such as ovulation and pregnancy. Certainly, the preclinical experiments can help us develop workable systems that can then be transferred to the clinical arena instead of *vice versa*, as has happened with this study of mammary carcinoma.

Acknowledgements

The authors wish to acknowledge the expertise of Ronald C. Wesley for construction of the apparatus and sensors illustrated in Figure 1. We also acknowledge the assistance of Thanh N. Nguyen for the drawings and Marilyn J. Rose for preparation of the manuscript.

REFERENCES

1 Burr HS, Smith GM, Strong LC: Bioelectric properties of cancer-resistant and cancer-susceptible mice. Am J Cancer 1938 (32):240-248

2 Langman L and Burr HS: A technique to aid in the detection of malignancy of the female genital tract. Am J Obstet Gynecol 1949 (57):274

3 Morris DM and Hirschowitz B: Electrical monitoring of breast cancer. J Bioelectricity 1982 (1):155-159

4 Nordenström BJW: Biologically Closed Circuits. Clinical, Experimental, and Theoretical Evidence for an Additional Circulatory System. Nordic Medical Publications, Stockholm, 1983

5 Faupel ML and Kornhauser S: Electromammography: A new modality for diagnosing breast cancer. Am J Electromed 1991 (22):88-91

6 Weiss BA, Ganepola AP, Freeman HP, Hsu YS, Faupel ML: Surface electrical potentials as a new modality in the diagnosis of breast lesions - A preliminary report. Breast Dis 1994 (7):91-98

7 Green EL (ed) Biology of the Laboratory Mouse. McGraw-Hill Book Company, New York 1966, 2nd Edition pp 189-196

8 Henry, RT and Casto R: Simple and inexpensive delivery of halogenated inhalation anaesthetics to rodents. Am J Physiol 1989 (257):R668-R671

Dedicated Systems for Surface Electropotential Evaluation in the Detection and Diagnosis of Neoplasia

Mark L. Faupel [1] and Yu-Sheng Hsu [2]

1   The University of Georgia, Boyd Graduate Studies, Research Center, Brooks Drive, Athens, GA 30602, U.S.A.
2   Georgia State University, Department of Math and Computer Science, 30 Prior Street, Atlanta, GA 30303, U.S.A.

The key to effective measurement and analysis of direct current (dc) skin potentials is absolute maintenance of signal integrity from the skin surface to the signal processing components of the computer's central processor [1]. This is critical because of the inherent low amplitude of biological dc potentials. At any point in the electronic path from skin sensor to device, potential exists for noise to intrude upon signal, thereby degrading diagnostically useful information. The central themes of this chapter are the design considerations of system components necessary for keeping noise to a minimum, and methods used to extract the critical diagnostic features from the resultant dc potentials.

System Components

The Sensor

The skin sensor is responsible for transmitting current from the skin through an electroconductive medium (ECM) to the sensor element, which transduces ionic conduction to metallic conduction. The resultant signal is then relayed to the computer's central processing unit (CPU) for processing. The design and performance characteristics for an effective dc sensor are different from those of electrodes designed for measuring alternating current (ac) signals such as those used with electrocardiography (ECG) and electroencephalography (EEG). For example, US national standards for single-use ECG electrodes allow the dc offset of a pair of electrodes (i.e., the spurious dc current gener-

ated by electrochemical interactions between electrode components) to be as high as 100 millivolts (ANSI/AAMI standard). Since effective use of dc signals for cancer diagnosis requires discrimination at the one millivolt level, standards for ECG electrodes are grossly excessive.

Effective analysis of dc signal measurement is tied to use of a sensor with low inherent dc offset. The key here is the interaction between the ECM, the sensor element, and the top snap of the sensor. Figure 1 shows an exploded view of a sensor designed specifically for dc measurements. The top snap (A) is conductive and connects to the cable which feeds the signal to the device. The liner (B) provides rigidity and strengthens the contact points between the top snap and the sensor element. The foam ring (C) is flexible enough for curved surfaces and helps maintain optimal spacing between sensors. The sensor element (D) transduces ionic signal to metallic signal. The reticulated sponge (E) provides an effective substrate for the ECM, which is essential for maintaining a conductive bridge between skin surface and sensor element.

The presence of any two dissimilar materials (especially if they are conductive) in an electrolytic environment is enough to produce dc current [2-5]. This phenomenon is often referred to as the battery effect or dc offset of an electrode. One only need recall the uncomfortable experience of inadvertently having a small piece of aluminum foil in contact with a silver tooth filling. The combination of silver and aluminum metals in a saliva bath produces the painful electrical charge. A similar phenomenon occurs in most conventional ECG electrodes.

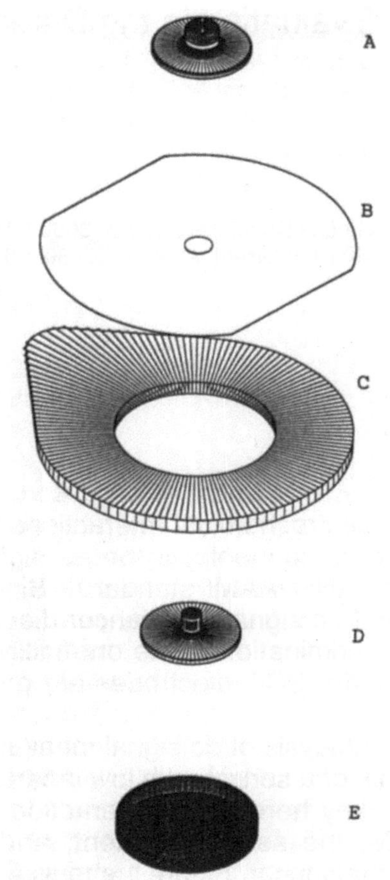

Fig. 1. Exploded view of direct current sensor system

The silver/silver chloride of the sensor element can react with the stainless steel or nickel plated brass of the top snap when in the presence of a liquid ECM. Although electrode manufacturers have developed techniques for reducing dc offset (for example, the development of nonliquid ECMs [6]), these innovations have come at the cost of increased system electrical impedance (and its dc analogue, resistance), which can be another source of noise.

The optimal single-use sensor system for biological dc measurement requires both low dc offset and low resistance. This can be accomplished by designing a sensor which utilizes the excellent conductive properties of a low-viscosity, high electrolyte-content ECM, along with a sensor which includes only one metallic component, in this case silver. Such a design circumvents the problem of a dissimilar metals reaction which can be exacerbated by use of low-viscosity, high electrolyte-content ECMs. Under strict manufacturing controls, sensors of this type can be made with dc offsets less than 0.5 mV and with very low impedance characteristics.

Ions from the skin surface are driven into the ECM by electromotive force, or the physical propensity of ionic concentrations to equilibrate. At the ECM/sensor interface, a charge gradient between ions in the ECM and the discharge of ions from the sensor is formed. This has been referred to as the electrical double layer and can be visualized as two parallel sheets of charge with opposite sign [7-9]. Maintenance of a stable double layer is an integral part of keeping noise to a minimum in dc measurement systems. Two ways of promoting a stable double layer are to isolate the sensor a short distance away from the skin surface and to keep patient movement to a minimum [10,11].

Another source of potential noise is resistance in a circuit, often referred to as Johnson noise, named after its discoverer [12]. Resistance in any circuit creates spurious wide-band voltages which result from the random motion of charge carriers in the conductor. Johnson noise can be ameliorated by using low resistance sensors and limiting the bandwidth of the measuring apparatus. Failure to control Johnson noise limits the level of minimum signal which is detectable [9].

Although numerous types of metallic electrodes have been used for recording biopotentials, sensors which utilize silver/silver chloride at the ECM/sensor interface appear to keep dc offsets to a minimum [13-15]. Care should be taken to ensure that the silver chloride finish is uniformly distributed and that impurities in the metals are minimal [1]. Likewise, impurities in ECM components also can lead to dc offset.

The Cable System

Once the signal is detected by the sensor, it is relayed to the CPU of the recording device. Cable system design is important in this regard, both to ensure that no signal degradation occurs along the signal path and that the design takes into account ergonomic factors. It is important that no leads be left disconnected from the patient, since open channels from disconnected sensors can function as antennae which pick up ambient electromagnetic noise. These spurious signals can then saturate other channels through crosstalk, which is the uncontrolled signal from one channel that impinges on an adjacent channel. The connection between the terminal ends of the cable must be secure, while also allowing for rotational freedom for adjusting the positioning of the leads with regard to the patient.

It is possible to test directly the circuit integrity of the cable and device by utilization of a field test calibration unit. This unit produces a series of known voltages via different resistor networks. Attaching the cable terminals to the prescribed outputs of the field test unit allows a check of the system using custom software built into the device. A final design consideration is to isolate the cable system from the power supplies (both line and internal) so that current cannot leak back into the patient.

The Device

A major design consideration for an effective dc measurement device is to make the input impedance many times higher than that produced by the skin and sensor interfaces. Failure to do so results in decreased signal amplitude and, most important, loss of low frequency (e.g., dc) information [16,17]. Sampling of dc potentials (e.g., number of samples and over what time period) ideally should be under computer control and it is here that microprocessor technology is utilized to its fullest. In digital systems, selection of the optimal sampling strategy is a balance between two factors: 1) acquiring enough samples per unit time to be representative of depolarization due to increased proliferation and 2) avoidance of taking samples over an extended time frame which might reflect dc drift. In other words, measuring the precise time slice of dc activity representative of cell proliferation is an important design consideration. It can be shown that reducing the number of dc samples per unit time can lead to irreproducible dc measurements.

Computer control of signal acquisition allows each sensor to be sampled via a multiplexed system. In a multiplexed system, each sensor's voltage is sampled many times in a precise sequence, using a single analogue-to-digital converter/amplifier. Because only one amplifier (rather than multiple amplifiers) is used, this system ensures that individual channels are calibrated with reference to each other. The resultant individual voltages are then averaged to provide a composite voltage for each sensor site. Since all individual samples can be stored and processed, digitally averaged voltages and variability in voltages over time can be analyzed for signal integrity. Comparisons in the averaged voltages are then used to identify areas of relative depolarization on the breast surfaces, as described in the chapters of this volume which cover clinical results.

The fastest multiplexed dc measurement system currently available for breast electropotential evaluation can scan up to 96 sensors 150 times over a 1.5-minute period. The array of sensors used in previous diagnostic studies [see Dickhaut et al., this volume] consists of 16 measuring sensors, which can be scanned 150 times each in about 18 seconds.

Pattern recognition of dc potentials is enhanced by effective filtering of the periodic electrical signals produced by cardiac and neural activity. These periodic signals typically range from about 1 Hz for cardiac signals to as much as 10 kHz for some neural signals. State-of-the-art digital filters are effective in keeping bandpass below 1 Hz. After filtering and averaging, dc voltages can be stored on disk, displayed on a CRT, or printed. Typically, averaged voltages corresponding to each sensor are displayed, along with two measures of variability. These two measures are the Modified Range (MRNG) and test Reliability (REL). These measures can provide important information regarding signal integrity. MRNG is the range in voltages from the 150 individual voltage samples. It is a modified range because device programming sets a limit of 20 mV (i.e., +/- 10 mV from the averaged voltage). If MRNG exceeds 20 mV, the individual voltages which exceed this value are filtered out and MRNG is recalculated. REL is expressed as a percentage of the 150 individual voltages and indicates how many individual voltages required filtering. High MRNG and low REL values may indicate that signal integrity has been compromised, perhaps by a sensor which has become disconnected from the patient.

One additional advantage of multiple sensor arrays is the richness of the data base which results from each patient test. As opposed to many other quantitative diagnostic and screening tests, such as serum assays, surface potential arrays provide multidimensional data. The key to effective diagnosis then becomes pattern recognition, which has become a science unto itself with the advent of sophisticated pattern recognition computer programmes. These techniques are described in the final section of this chapter.

Test Procedure

A primary advantage of surface electropotential measurement is that it is a non-invasive procedure. For breast evaluation, patients are tested in a recumbent position and are encouraged to relax. In some cases, patients have fallen asleep during testing. When explaining the test procedure to prospective patients, it is important to emphasize that no radiation or electrical current is being put into the body. The only discomfort reported by some patients has been during the removal of the adhesive sensors after the test, the degree of which is almost always reported as mild.

During clinical trials, the patient and her physician are masked from the results of the test to ensure that this experimental procedure can have no bearing on subsequent patient management. The technician should carefully explain all steps in the testing procedure and answer any questions the patient might have about the procedure.

Sensor Array Placement

Once the patient has given informed consent to be enrolled in the study, she is requested to lie down on the examining table and relax. The next step entails placement of sensors, which is done using a standard set of placement rules. For the diagnostic test*, in which a suspicious lesion has already been localized, the following set of sensor placement rules is carried out. First, the position of the suspicious lesion is identified, either by palpation or, if the lesion is not palpable, by extrapolation from a mammographic image. One sensor (marked LC or RC, depending on whether the lesion is in

the left or right breast) is placed over the centre of the lesion. Four additional sensors (upper, outer, lower, inner) are placed just outside the margins of the lesion in a north-south/east-west pattern. Minimal spacing between sensors should be no less than 3.3 cm apart, measuring from centre to centre of the sensor. If the lesion is very large, then sensors should be spaced further apart, such that the four sensors which surround the centre sensor are placed just beyond the margins of the lesion. Sensors can be placed off the breast tissue if necessary to maintain proper spacing.

After the array of five sensors is placed in the vicinity of the lesion, two additional sensors are placed in the centre of the two quadrants directly adjacent to the quadrant containing the suspicious lesion. An eighth sensor is placed in the axilla, where one would palpate for lymph node abnormalities. This pattern of sensor placement is then replicated in a mirror-image fashion on the opposite breast and axilla. The last step is to place a reference sensor on the thenar eminence (palm) of each hand. All potentials measured from both breasts and axillae are made relative to the palm sensors. Figure 2 shows a schematic representation of sensor placement for a patient with a previously localized lesion in the upper outer quadrant of the right breast.

Although well designed electropotential devices filter out most neuromuscular noise, additional steps can be taken to keep noise to a minimum. First, care should be taken to ensure that all sensors are firmly attached to the skin surface. An unattached or partially attached sensor can function as an antenna which picks up extraneous electromagnetic noise. A relaxed patient generally produces cleaner dc readings than an agitated one, although digital filtering can reduce these effects to a great degree. On the other hand, a patient who is constantly in motion or disturbing the palm sensors will disrupt the electrical double layer at the sensor interface. Such motion artefacts can lead to significant noise in the circuit. Fortunately, the device can be programmed to display the results of excessive noise and alert the technician that a retest is indicated.

The duration of the entire test procedure is determined primarily by three factors: the time it takes to apply the sensors, the time to reach equilibrium between the skin and ECM, and the

---

* An alternate sensor array, which would be more appropriate for screening asymptomatic patients, is discussed by Crowe and Faupel in this volume.

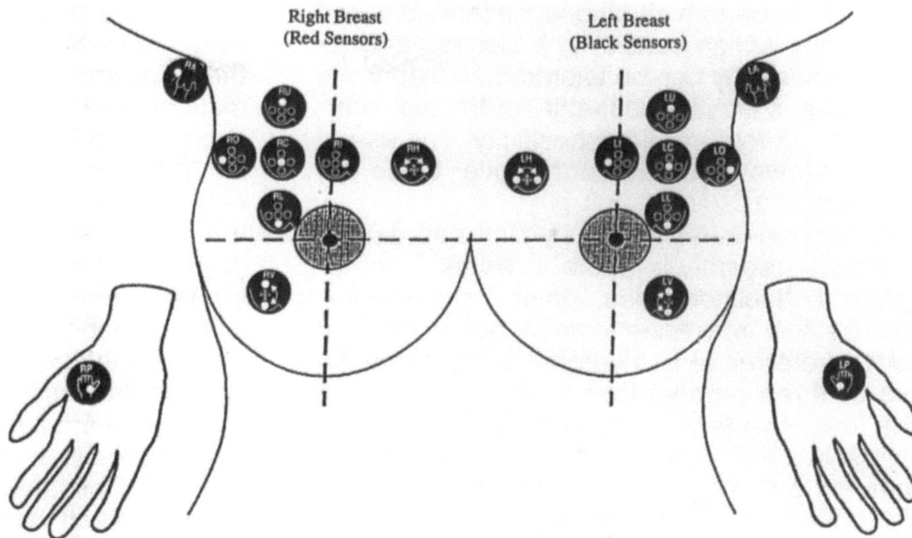

Fig. 2. Diagnostic sensor array (lesion in upper outer quadrant of right breast)

time it takes for skin surface potentials to be measured. A trained technician can apply the diagnostic sensor array within a few minutes. Depending on the type of ECM used, the time it takes for the ECM to penetrate the stratum corneum and produce stable readings can vary from 3 to 10 minutes. For clinical trials, the computer's internal timer is set so that readings are taken no less than 10 minutes after sensors are placed on the skin. This 10-minute period is then used for completing study documentation. The measurement period is relatively short; initial prototypes took about 3 minutes to run 2 successive diagnostic tests (one for each reference channel). The current version of the device can accomplish the same in about 36 seconds. All results are automatically saved on disk and can be printed immediately after the test.

Signal Processing

Because concurrent measurements are made from a multiplicity of sites, surface potential analysis provides information across several dimensions. Differentials can be calculated from the entire breast or specific regions within the breast. Potential differences can also be compared between the two breasts (the between-breast differential), or between corresponding pairs of sensors (mirror-site differential). This wealth of quantitative information presents both an opportunity and a challenge to ana-

lysts whose primary goal is to extract signals from an inherently noisy biological environment. The classic approach to this kind of pattern recognition problem is to employ linear discriminant analysis. This can be effective in establishing a set of parameters which predict a given state, such as malignant vs. benign disease. Its predictive value decreases if there are thresholds in the data. If thresholds in the data do exist, then non-linear discriminant methods are called for. One recently developed approach is tree analysis, such as CART [18], PIMPLE [19], MARS [20], and SUPPORT [21]. Decision trees, such as those produced by CART (Classification and Regression Trees) lead to a binary outcome (e.g. cancer vs. benign states). Each decision point in the tree is referred to as a node in which optimal cutoff values are established based on variables considered for the study. CART arrives at the best decision tree for a data set first by constructing a very large tree using certain optimal criteria, and then pruning back the branches to reveal the key predictive variables. CART trees can be constructed with two parameters which can be controlled by the user. One parameter is tree complexity, which is used to penalize a non-significant large tree, which may not be predictive of future data sets. The other parameter is the cost ratio between false-negative and false-positive results, which in essence allows the user to determine the trade-off between test sensitivity and specificity. The ability to trade sensitivity against specificity is a powerful feature of electropotential analysis. In certain situations, such as disease screening, a high sensitivity is desired and reduced specificity can be tolerated. In diagnostic situations, it may be desirable for the test characteristic to feature high specificity. The use of CART allows objective control over these parameters.

Another non-linear method which has gained popularity recently is neural networks. As opposed to decision trees, which function as a process flow from node to node, neural network nodes (referred to as neurons) are arranged in two or three parallel levels. Information from one level can feed back to influence decision points at other levels in the network, until a complex path of decisions is made which can discriminate between disease states.

Regardless of which pattern recognition strategy is adopted, assessing the predictiveness of the decision matrix for new populations is essential. The traditional approach has been to collect data for a subgroup of patients, develop the decision matrix, and then validate the matrix on a new group of patients. The former group is referred to as the training data set and the latter group is referred to as the test data set. The decision matrix developed for the training data set is cross-validated for the test data set. In this way cross-validation provides an index of how well the test sample will predict with regard to the population for which the test method was developed. Unfortunately, this strategy wastes resources by requiring the study to be run twice. In most studies data are a precious resource, and it is advantageous to use the largest possible population for development of the decision matrix [22]. Therefore, statisticians have developed many data resampling methods which utilize the same sample for decision matrix formation and validation. The first resampling techniques, referred to as the jackknife and resampling cross-validation, have been studied for several years. A newer resampling method, called the bootstrap [23], has been one of the most intensively researched methods in statistics during the past 15 years. The method consists of treating the existing sample as the population. One uses a computer to draw many samples from the test population in order to measure the bias between the sample and the population. Resampling cross-validation calls for reiteratively sampling random segments of the data to be used for decision matrix formation, with residual segments

being used for validation. For example, a decision matrix is constructed on a 90% sample of the population and then validated on the remaining 10%. Next, a different 90% sample is chosen at random and validated on the remaining 10%. The process is done recursively for numerous 90/10 splits of the data set. The total error generated by all the recursive samplings of the data base can be used as a surrogate for traditional validation techniques. We recently developed add-on bootstrap and cross-validation modules for CART, which allow independent assessment of test sensitivity and specificity sample bias.

Once a decision matrix for disease diagnosis has been validated, it can be programmed into the device software. This allows a test result to occur in real time. For each patient test, output from the device is anticipated to include averaged potentials from each sensor (along with MRNG and REL data), as well as a probability estimate as to whether the breast electropotential readings indicate the existence of malignancy. The physician would then use this information within the context of the patient's total medical profile, which includes results from other tests and medical history.

If this technology is adopted, it is anticipated that patients will undergo surface electropotential evaluation at regular intervals, similar to those recommended by various mammography screening programmes. This was simulated in preclinical studies (see the chapter by Long et al. in this volume) in which surface potential measurements were made several times in BALB/c mice both before and after injection of tumour cells. In this case, differences in the baseline potential differentials could be compared with post-injection potentials for each animal. To do this, the usual 2-sample t-test may not work well, since the post-tumour injection readings may be taken over a limited time period. In other words, the usual t-test may not detect a change in the mean of the post-injection readings when maximum deviations due to cell proliferation occur at only one or two points during a sequence of tests. Thus the aim is to develop a statistical method which can identify significant deviations (known as outliers) from a series of tests. There are many methods which can be applied to detect outliers, such as time series analysis or CUSUM method. However, most of these methods are restricted to analysing a single long series of data. Furthermore, these methods are unable to assess change patterns which may differ across subjects. We developed the Max-t method for the situation in which data are collected over a short series of tests with unequal temporal spacing. This method uses the maximum readings in the post-injection data combining with the baseline data set to form t-type random variables. The Komogorov-Smirnov test is then used to test for the significance of the outliers. Power studies via simulation method indicated that Max-t is superior to the 2-sample t-test when the size of the change is significant.

In addition to assessing tumour models in rodents, Max-t analysis of skin surface potentials would be effective for monitoring response to therapy. Autoregressive models also can be used in situations in which departures from baseline conditions need to be identified, and can be used even though data are unequally spaced in time or consist of relatively short sequences [24].

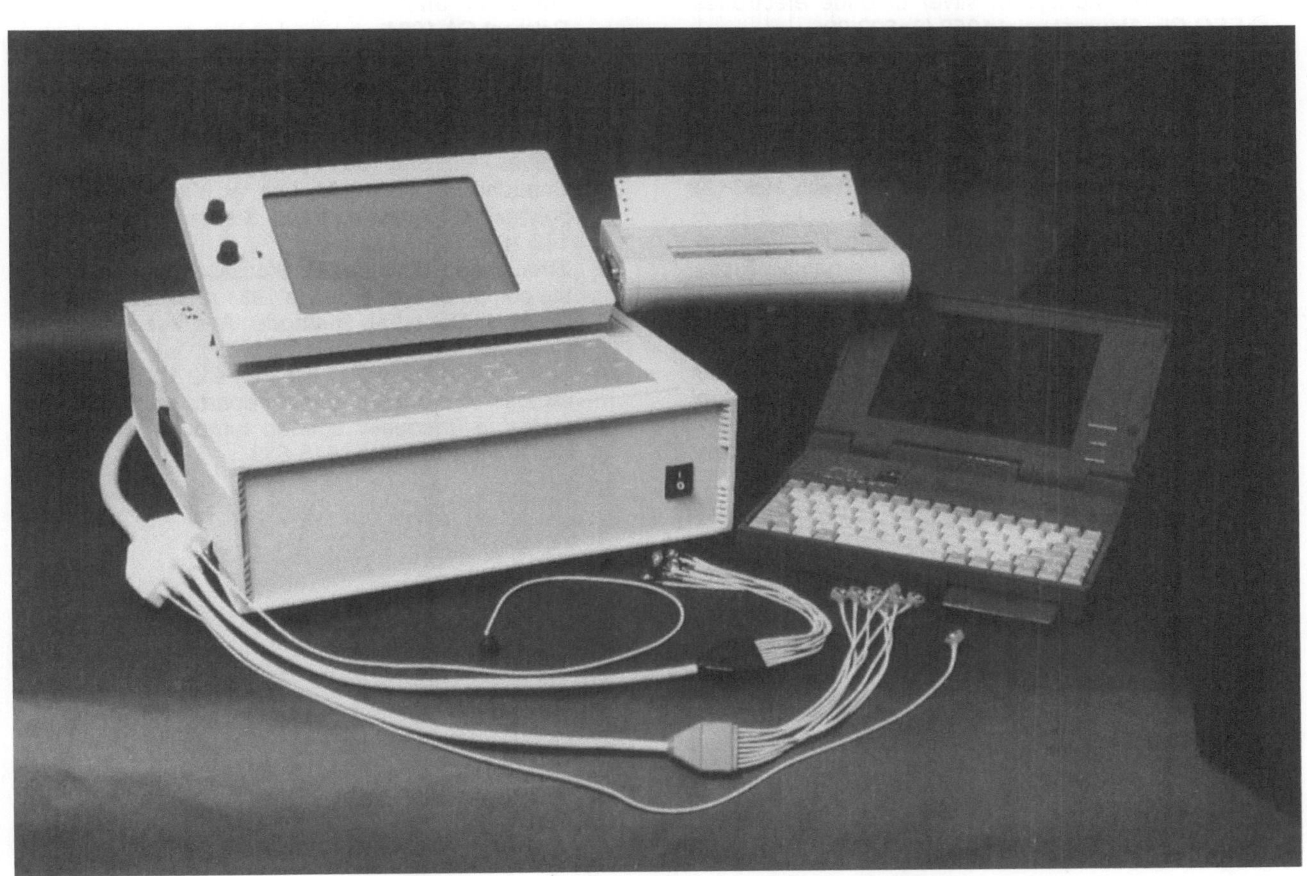

Biofield M4 Breast Biophysical Evaluation (BBE) Diagnostic Device. The system consists of patient cables (colour-coded red for right breast and black for left breast), the measurement device with LCD display, printer, and laptop computer for data storage

REFERENCES

1   Geddes LA and Aronson S: Electrode potential stability. IEEE Transactions on Biomedical Engineering 1985 (32):987-989

2   Lykken DT: Properties of electrodes used in electrodermal measurement. J Comp Physiol Psych 1959 (52):629-634

3   O'Connell DN, Tursky B, Orne MT: Electrodes for recording skin potential. Arch Gen Psychiat 1960 (3):252-258

4   Goldstein AG, Sloboda W, Jennings JB: Spontaneous electrical activity of three types of silver EEG electrodes. Psychophysiol Newsletter 1962 (8):10-16

5   Cooper R: Storage of silver chloride electrodes. EEG Clin Neurophysiol 1956 (8):692-696

6   Roman JA and Lamb LE: Electrocardiography in flight. Aero Med 1962 (33):527-544

7   Parsons R: Electrode double layer. In: CA Hampel (ed) The Encyclopedia of Electrochemistry. Reinhold Publishing Co, New York 1964

8   Cooper R: Electrodes. Amer J EEG Tech 1963 (3): 91-101

9   MacFarlane PW and Veitch Lawrie TD: Comprehensive Electrocardiology Vol 1. Pergamon Press, Oxford 1972

10 Atkins AR: Measuring heart rate of an active athlete. Electron Eng 1961 (33):457-468

11 Kahn A: Motion artifacts and streaming potentials in relation to biological electrodes. Digest 6th Int Conf Med Electr Biol Eng 1965 (6):562-563

12 Johnson JB: Thermal agitation of electricity in conductors. Phys Rev 1928 (32):97-109

13 Geddes LA and Baker LE: Chlorided silver electrodes. Med Res Eng 1967 (6):33-34

14 Lucchina GG and Phipps CG: A vectorcardiographic lead system and physiologic electrode configuration for dynamic readout. Aero Med 1962 (33):722-729

15 Lucchina GG and Phipps CG: An improved electrode for physiological recording. Aero Med 1963 (34):230-231

16 Geddes LA, Baker LE, McGoodwin B: The relationship between electrode area and amplifier impedance in recording muscle action potentials. Med Biol Eng 1967 (5):561-568

17 Geddes LA and Baker LE: The relationship between input impedance and electrode area in recording the ECG. Med Electr Biol Eng 1966 (4):439-450

18 Breiman L, Friedman JH, Olshen R, Stone CJ: Classification and Regression Trees. Wadsworth, Belmont CA 1984

19 Breiman L: The method of estimating multivariate functions from noisy data. Technomet 1991 (33): 125-160

20 Friedman JH: Multivariate adaptive regression splines. Ann Stat 1991 (19):1-14

21 Chaudhuri P, Huang M, Loh W, Yao R: Piecewise polynomial regression trees. Statist Sinica 1994 (4): 143-167

22 Efron B and Tibshirani R: Statistical data analysis in the computer age. Science 1991 (253):390-395

23 Efron B: Bootstrap methods: Another look at the jackknife. Ann Stat 1979 (7):1-26

24 Schlain BR, Lavin PT, Hayden CL: Using an autoregressive model to detect departures from steady states in unequally spaced tumour marker data. Stat in Med 1992 (11):515-532

Is There a Need for a New Detection Technique for Breast Cancer?

Volker Barth and Johannes Herrmann

The Central Radiological Institute, Hirschland Strasse 97, 7300 Esslingen am Neckar, Germany

In West Germany, 25,000 women every year are diagnosed as having breast cancer, and 13,000 die from this disease. One out of every 13 women will be diagnosed with this cancer in her lifetime.

If screening mammography is integrated into the national health plan and covered by national health insurance, it is estimated that the mortality rate from breast cancer could be decreased by 30%. Problems with implementing a screening programme include limited financial resources of the national health plan and quality control of individual mammograms. The combined results of 7 different screening programmes in Europe indicate that up to 60% of all tumours diagnosed were at stage T1. Until recently, in Germany the early diagnosis of breast cancer has been based on clinical palpation by a physician and educating patients in breast self-examination [1].

With a compliance rate of only 31%, 2,989 malignant breast tumours have been diagnosed and documented per year. This indicates that only 12% of the expected incidence was diagnosed by mammographic screening. Considering the results of the above-mentioned European 7-centre screening programme, it is obvious that this is not adequate [1].

The aim of breast cancer screening is to detect invasive carcinoma at an early stage, i.e., smaller than 1 cm in size and without axillary lymph-node involvement, as well as to be able to detect non-invasive carcinoma, which may have a latent period of 5 to 15 years, during which time it may become invasive. Early and pre-invasive carcinoma can be diagnosed mammographically in 60% of cases due to suspicious microcalcifications. Unfortunately, these microcalcifications can also be associated with benign changes. Up to two thirds of all mammographically suspicious microcalcifications are benign on histological evaluation. This means that as many as 2 out of every 3 women undergo unnecessary open biopsy to resolve mammographically identified microcalcifications [2]. Thus in some instances mammography cannot differentiate between malignant and benign calcifications. Early and pre-invasive carcinomas which do not present with calcifications are discovered by finding a palpable mass in the breast or by chance through biopsy of a benign change in the breast.

In many cases, early and pre-invasive carcinomas in the breast remain undetected because mammography of radiographically dense breasts characteristic of younger women does not identify them. Because of this, it would be desirable to have a simple and cost-effective test adjunctive to mammography which can further differentiate between malignant and benign diseases.

Mammography has a high sensitivity (80%) but unfortunately a relatively low specificity (approximately 50-60%). This means that many changes, for example calcifications, stellate opacities, asymmetry, architectural changes and round mass lesions are discovered for which the predictive value of a malignant process differs considerably. For example, round mass lesions have a predictive value of about 5%, while stellate opacities have a predictive value of about 85% [2].

The German Mammographical Breast Cancer Study reports a general positive predictive value of mammography of 28% for carcinomas [1]. This means that for every 100 women undergoing surgery, 30 have a malignant lesion and 70 have a benign process. It is clear that this ratio is unacceptable, especially considering the physical and psychological trauma of unnecessary surgery, as well as the high cost incurred by additional diagnostic workup of mammographically suspicious lesions.

Is there a method that could help reduce unnecessary and costly diagnostic procedures? A new test must be found which would increase the positive predictive value of screening to 50-60%.

Because of its low specificity, the expense and exposure to radiation, mammography is not an ideal screening method. Some additional criticisms of mammography include its limitations in younger women, its subjective and often difficult interpretation, and the lack of reliable data regarding the best interval between serial examinations [2]. Thus the challenge is to develop a test which can increase the effectiveness of mammography and thereby lower the frequency of additional, expensive diagnostic workup, including unnecessary excisional biopsies.

What Methods are Available for the Diagnosis of Breast Cancer?

Due to the low positive predictive value of screening mammography, additional diagnostic methods are needed to differentiate radiographically ambiguous changes. This is not a trivial matter, since it is expected that between 5% and 7% of all screened women will require additional diagnostic workup [3]. The additional methods which are available for post-screening diagnostic workup are adjunctive to both mammography and physical examination (palpation), and can be divided into 2 broad categories:

1) Imaging modalities (e.g., ultrasound, further mammography including magnification mammography, computer tomography, and magnetic resonance imaging (e.g., nuclear spin tomography)).
2) Tissue sampling modalities (e.g., fine-needle aspiration cytology, stereotactic fine-needle aspiration cytology, stereotactic core-needle biopsy).

Tissue sampling methods such as fine-needle aspiration cytology are relatively well established and have been described elsewhere in detail [3,4-8]. In this chapter, we will concentrate on some of the emerging imaging methods

which are adjunctive to screening mammography.

Inspection and palpation of the breast and its draining lymph nodes can be of value because up to 10% of all palpable breast carcinomas are not detected by mammography [2], thus palpation increases the overall diagnostic sensitivity. However, the positive predictive value of palpable changes in the breast is reported to be only about 8% [9]. Sensitivity of palpation as a diagnostic method is reduced for smaller lesions and in younger women [10,11]. Because of this level of diagnostic inaccuracy, in the Netherlands and Sweden palpation is no longer used as a supplement to screening mammography.

Adjunctive Imaging Modalities

Ultrasound

This non-invasive procedure is effective for differentiating solid from cystic lesions [12-14]. The problem is that certain lesions such as fibroadenomas, cellular neoplasms, lymphomas and mucinous carcinomas can also produce images characteristic of cysts, leading to false-negative results. Ultrasound is not appropriate as a method for early detection of occult carcinomas. Microcalcifications, a typical finding associated with early carcinomas, are not evident on ultrasound examination [14]. Although ultrasound can differentiate solid from cystic masses, other methods such as fine-needle aspiration cytology (FNAC) are as effective for this purpose and are more economical in cost and time [12]. Sonograms appear to be most useful in those cases where FNAC is not feasible, either due to difficulty in locating the exact position of the mass, or in cases where the patient finds aspiration objectionable [13,14]. Ultrasound cannot reliably distinguish malignant from benign lesions and is not sufficiently sensitive for lesions less than 1 centimetre in size [12,13].

Diagnostic mammography

Additional diagnostically useful information can be gained from using multiple mammographic projections or increased magnification. Multiple projections indicate the margin characteristics of lesions, while magnified views increase absolute resolution, which increases the probability of identifying spiculations and microcalcifications associated with carcinomas [2]. The inherent limitations of mammography cannot be overcome completely by multiple or magnified views. Many benign conditions, such as fat necrosis, scars, and fibrocystic changes, share anatomical characteristics associated with carcinoma, and in some cases cancers have regular margins and lack evidence of microcalcifications [12].

Computer tomography (CT)

This technique facilitates the differentiation of density variations but does not equal the ability of mammography to spatially pinpoint the location of a lesion. Furthermore, the sensitivity of computer tomography has been reported to be about 70% [15]; CT is also costly and involves high doses of radiation [12]. In some cases, contrast agents can be used to increase resolution, as in the more recently developed spiral technique, which can image tumours ranging in size from 0.6 to 1.0 cm. CT is only rarely considered of any value over conventional mammography, namely in the examination of young, radiologically dense breasts and in the evaluation of fibrocystic mastopathy. However, it may be of value in treated breasts.

Magnetic resonance imaging

The most sensitive examination in the localization of breast cancer today is nuclear spin tomography [16-18]. Using radiowaves within a magnetic field (i.e., without radiation), cross-sections can be taken in any spatial dimension of the human body. For examination of the breast, the patient lies on her stomach and the breast is placed into a plastic bowl surrounded by a double spool magnet. Nuclear spin tomography is enhanced by injection of Gadolinium (a paramagnetic substance), which functions as a contrast agent. Gadolinium tends to accumulate in proliferating glandular tissue, thereby intensifying the signal properties. This enriched contrast method may be useful for:

1. Differentiation of smooth, solid, well defined nodules [16,18].
2. Examination of the radiologically dense breast [16].

3. In known carcinoma, to identify uni vs multifocality for breast conservation [16,18].
4. The examination of the healthy breast.

It also enables examination of the chest wall and axillary regions. It can detect changes associated with pre-carcinoma, differentiating those atypical lesions as well as carcinoma *in situ* and invasive tumour stages, which previously was not possible.

If neither breast shows Gadolinium accumulation, it is likely that no proliferating process larger than 5 mm is present. Pre-invasive carcinoma as well as certain types of invasive carcinoma, e.g., invasive tubular carcinoma, show only a small degree of contrast concentration.

High cost is the primary problem of nuclear spin tomography. In Germany one examination costs 4 to 5 times more than a mammogram. Additional problems are:

1. Patients suffering from claustrophobia (7%) are not able to lie still within a closed magnet.
2. It is difficult to mark suspicious lesions that should be removed surgically.
3. The positive predictive value appears to be no better than that of mammography [12,19].

Thus the primary advantage of this modality is a high sensitivity for identifying proliferative processes within the breast. The ideal test for the detection of cancer should maintain this high sensitivity, but at lower cost and with more universal clinical acceptance.

Summary

Physical examination, mammography, ultrasound, fine-needle aspiration, computer tomography, nuclear spin tomography, and histological examination are of great importance at different stages in the diagnosis of benign and malignant breast tumours. Because the value of diagnostic examination is measured by the ability of the test to detect early cancer at the occult stage, all examinations should be evaluated based on this ability.

The diagnostic and therapeutic goal in preventative breast care is to diagnose early malignant changes to improve the prognosis of the patient. Mammography is currently the examination of choice as an initial screening method to detect clinically occult carcinomas. The limitations of mammography include:

1. It is technically difficult and somewhat expensive to perform.
2. It is difficult to interpret and has a high false-positive rate.
3. It is useful only in women older than 40 years, with some authorities claiming it to be effective only in women older than 50 years.
4. The best interval between serial examinations is still unknown.
5. Patients suffer anxiety due to exposure to radiation.

Of the imaging examinations, only nuclear spin tomography can be considered as more sensitive for the detection of breast cancer than mammography. However, it is expensive and, like ultrasound, appears unable to detect small but important microcalcifications. Therefore, neither ultrasound nor nuclear spin tomography are currently suitable for the early diagnosis of breast cancer.

Do Clinicians Need Another Method to Detect Breast Cancer?

An alternative method to detect breast cancer could be the Breast Biophysical Examination (BBE), as described in this publication. With clinical examination and palpation having low effectiveness, and with the costs and other limitations of mammography and nuclear spin tomography as outlined above, a gap exists in the diagnostic pathway currently in use.

It would be advantageous to have a method which could be used immediately after the initial screening process that could identify an abnormal proliferative or atypical change in the breast. This appears to be the case with the BBE, which measures the shift in electrical field associated with carcinomas and epithelial proliferation. A major advantage is that these electrical field shifts can be detected non-invasively at the skin surface using specially designed sensors.

Because the characteristics of this test address many of the disadvantages of currently employed modalities, a working group consisting of two clinics in Germany (the author's group in Esslingen, as well as Frischbier/Schreer in Hamburg) recently began assessing this new technology. The results of 158 cases, including 87 known cancers and 71 known benign lesions, are discussed in the clinical results section of this volume. The tested lesions, 56 of which were non-palpable, included both invasive and non-invasive carcinomas, as well as a variety of benign lesions including atypical hyperplasia, intraductal papilloma, fibrocystic changes, fibroadenoma, and adenosis. The highest surface electrical potential differences were shown by malignant lesions in symptomatic breasts. Non-invasive carcinomas produced higher breast potential differences than invasive carcinomas, although the difference was not statistically significant. A retrospective analysis of the data allowed the formulation of a decision-making algorithm. This algorithm revealed a sensitivity of 94.3% and a specificity of 77.5%. These preliminary results are encouraging, and a multicentre double-blind study is under way to prospectively verify these results. If the preliminary results can be confirmed, BBE would be an ideal post-screening test for the following reasons:

1. Women of any age could be examined as often as deemed necessary, and radiation exposure is not a concern. This is important since breast cancer does affect women under the age of 40. Mammography is not indicated for this age group, unless clinical suspicion exists.
2. BBE is technically easy to perform, and could be employed in all medical practices. Because infrastructural requirements are minimal, it is likely to be less expensive than nuclear spin tomography and mammography.
3. If the preliminary results are confirmed in the blinded multicentre study, high-risk groups could be identified and closely followed. These groups include women with:
   a) a family history of breast cancer in the patient's mother or sister at an early age;
   b) patients with a previously identified breast cancer;
   c) histologically proven precancerous changes.

Currently, the above groups must have mammography as well as additional diagnostic workup. The BBE would be preferable because it appears to be able to distinguish those patients who actually require additional, expensive diagnostic workup from those patients who simply need to return for routine follow-up.

Conclusions

A variety of diagnostic tests can be used to identify early stage malignant tumours. Each of these tests, including clinical evaluation and mammography, play a valuable role in prevention programmes. Mammography, the most effective screening method, is expensive and has a low specificity, resulting in unnecessary surgical biopsies with associated costs. The issue of radiation exposure negatively affects general acceptance, especially in younger women. Ultrasound, although less expensive, is not suitable for early diagnosis due to its failure to detect small microcalcifications. Nuclear spin tomography has similar limitations and the expense associated with this test limits its practical use.

In the spectrum of diagnostic techniques, a method is urgently needed that would be non-invasive, economical, and have the ability to differentiate between benign and malignant changes. The preliminary results indicate that this gap could be filled by the BBE. A large multicentre, double blind-study must prove if these criteria are objectively met by this technology.

REFERENCES

1   Frischbier HJ, Hoeffken W, Robra BP: Mammographie in der Krebsfrüherkennung. Ergebnisse der Deutschen Mammographie-Studie. Enke-Verlag, Stuttgart 1994
2   Barth V: Mammographie: Intensivkurs und Atlas für Fortgeschrittene. Enke Verlag, Stuttgart 1994
3   Tubiana M, Holland R, Kopans DB et al: Advisory Report of the European School of Oncology to the Commission of the European Communities "Europe Against Cancer" Programme. Advisory Report 1993
4   Dowlatshahi K, Yaremko ML, Kluskens LF, Jokich PM: Nonpalpable breast lesions: Findings of stereotaxic core-needle biopsy and fine needle aspiration cytology. Radiology 1991 (181):745-750
5   Elvecrog EL, Lechner MC, Nelson MT: Nonpalpable breast lesions: Correlation of stereotaxic large-core needle biopsy and surgical biopsy results. Radiology 1993 (188):453-455
6   Evans PW: Fine-needle aspiration cytology and core biopsy of non-palpable breast lesions. Curr Op Radiol 1992 (4):130-138
7   Minkowitz S, Moskowitz M, Khafif RA, Alderete MN: Tru-cut needle biopsy of the breast: An analysis of its specificity and sensitivity. Cancer 1986 (57):320-323
8   Malley F, Casey TT, Winfield AC, Rodgers WH, Sawyers J, Page DL: Clinical correlates of false negative fine needle aspirations of the breast in a consecutive series of 1,005 patients. Surg Gyn Obstet 1993 (176):360-364
9   Thomas BA and Price JL: The place of clinical examinations in breast cancer screening. Excerpta Medica, Amsterdam 1989:11-24
10 Cardona G, Cataliotta L, Ciatto S, Rosselli Del Turco M: Reasons for failure of physical examination in breast cancer detection (analysis of 232 false-negative cases). Tumori 1983 (69):531-537
11 Ciatto S, Rosselli del Turco M, Catarzi S, Cataliotti L, Cardona G, Teglia C, Pacini P, Caridi G: Causes of breast cancer misdiagnosis at physical examination. Neoplasm 1991 (38):523-531
12 Donegan WL: Evaluation of a palpable breast mass. New Engl J Med 1992 (327):937-942
13 Rozner D and Blaird D: What ultrasonography can tell in breast masses that mammography and physical examination cannot. J Surg Oncol 1985 (28):308-313
14 Sickles EA, Filly FA, Callen PW: Breast cancer detection with sonography and mammography: comparison using state-of-the-art equipment. AJR 1983 (140):843-845
15 Karsell PR: Computerized tomographic mammography. JNCI 1976 (5):8-12
16 Harms SE, Flamig DP, Hesley KL et al: Fat suppressed three-dimensional MR imaging of the breast. Radiographics 1993 (13):247-267
17 Hickman PF, Moore NR, Shepstone BJ: The indeterminate breast mass: assessment using contrast enhanced magnetic resonance imaging. Br J Radiol 1994 (67):14-20
18 Heywang-Kobrunner SH: Contrast-enhanced magnetic resonance imaging of the breast. Prog Clin Radiol 1994 (29):94-104
19 Khalkhali I, Mena I, Diggles L: Review of imaging techniques for the diagnosis of breast cancer: a new role of prone scintimammography using technetium-99m sestamibi. Eur J Nucl Med 1994 (21):357-362

Breast Biophysical Examination (BBE): Potential Applications in Clinical Practice

Mirella Merson [1] and Bruce A. Bach [2]

1   Department of Surgical Oncology, Istituto Nazionale Tumori, Via Venezian 1, 20133 Milano, Italy
2   Assistant Clinical Professor of Medicine, University of California, 505 Parnassus Avenue, San Francisco, CA 94143, U.S.A.

The early detection of breast cancer and the prompt selection of appropriate therapy (generally involving a combination of modalities) have as their goal the reduction of the morbidity and mortality of this disease. Current opinion, based on large prospective studies, suggests that this aim can be achieved [1-6].

The purpose of this chapter is to explain from a clinical perspective how the measurement of surface electropotentials may contribute to the various stages of breast cancer management (Fig. 1). These stages are 1) screening, 2) diagnostic evaluation, and 3) staging with treatment selection. Table 1 summarizes the characteristics of Breast Biophysical Examination (BBE) which offers unique potential advantages.

Possible applications of the BBE are considered in the order in which they might be used in a typical case: in screening, in diagnostic evaluation, in staging and treatment selection, and during follow-up for local recurrence. Further applications may derive from the relationship demonstrated between BBE and the intensity of cellular proliferative activity. This gives the technique additional potential as a prognostic tool and a possible role in monitoring the response to systemic treatment.

Screening

Goals and Limitations

The importance of early diagnosis in improving survival in breast cancer is well recognized [1]. The goal of breast cancer screening is to detect the presence of disease at an early stage in asymptomatic women. The mainstay of current screening programmes is X-ray mammography. There is little doubt that mass screening programmes which have been sufficiently accepted by women over 50 years of age have been successful in reducing the size and extent of breast cancers diagnosed [2].

Benefit from screening programmes has been more pronounced for women over 50 than in younger groups. Although mammography is able to detect non-palpable lesions, it is less effective in younger women whose higher levels of oestrogen result in radiographically dense breasts [2,3]. The technique is therefore less sensitive in picking up breast cancer in this population. This observation, which has been confirmed repeatedly, appears to limit the utility of current mammographic techniques in detecting the presence of early-stage asymptomatic breast cancer in younger women [7].

Table 1. Characteristics of BBE

The procedure

1)   Is non-invasive
2)   Objectively detects the degree of proliferative activity *in vivo*
3)   Has a sensitivity of 97% and a specificity of 81% (see Dickhaut et al., this volume), independent of the age of the woman, the size of the lesion, and previous local treatment
4)   Can be repeated an unlimited number of times
5)   Provides immediate objective results in the care setting

---

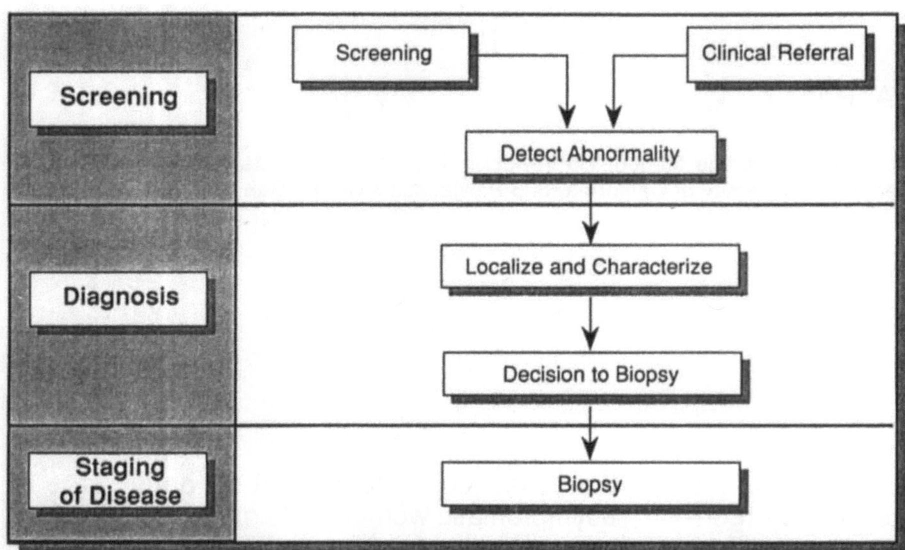

Fig. 1. The three-stage model of breast disease management showing objectives by stage

In addition to age-related sensitivity, given the wide range of frequently subtle changes in radiographic appearance, the interpretation of mammograms is subjective and reader dependent. Diagnostic accuracy depends on the experience and training of those involved in interpreting the films and performance is variable between centres and individuals [8].

The desire to increase the yield of early disease detection using mammography in populations over 50 has led to 5-10% of women being recalled for additional studies following the detection of a possible abnormality on screening. A challenge for screening programmes is the fact that many of these radiographically detected lesions are non-palpable or difficult to palpate and uncertainty in image interpretations requires further investigation. In parallel with mammography, breast self-examination yields a population of women with palpable abnormalities of uncertain significance [7] (Fig. 1).

A Role for BBE

In essence, breast biophysical examination involves the recording of low-amplitude electropotential "signals" produced by upregulated proliferation in epithelial cells. The technique may be particularly helpful in women with mammographically suspicious but non-palpable lesions. A positive BBE might reveal on the skin surface whether a lesion is likely to be ma-

lignant. Women with a suspicious radiographic finding and a BBE suggestive of "epithelium at risk" could then be selected for rapid evaluation using conventional diagnostic techniques.

Non-palpable lesions may be difficult to visualize against a background of dense breast tissue in young women's breasts and to date there has been no way of detecting early lesions that are not revealed by mammography, particularly if they are deep in the breast parenchyma. The potential sensitivity of mammography for early lesion detection is thus compromised in younger women. Given that such radiographically silent lesions are now not detected until they form localized masses, it is possible that BBE may play a role in identification of areas of abnormality which require further assessment.

Using a whole breast array of sensors more widely dispersed than those employed for localized lesions may permit a biophysical examination of the entire breast [see Crowe and Faupel, this volume]. This would give BBE a potential role in screening programmes. If electropotential measurement could be used to detect radiographically silent or less prominent lesions, an additional population of women with early disease, which is potentially curable, might be identified.

However, the major role of BBE might be in the assessment of women referred for workup of possible breast cancer, given the relatively low prevalence of breast cancer in screening populations. This would take advantage of the very high negative predictive value of the BBE which, based on a sensitivity of 97% and a specificity of 81%, would be 99.6% in a low-prevalence post-screening population. This would imply that women with a negative BBE are unlikely to have breast cancer.

Diagnosis

Current patients selected for diagnostic workup come from two sources: those identified as part of a formal breast cancer screening programme and those who seek medical attention for breast symptoms once a potential breast lesion has been identified. The clinician's aim at this point is to localize and characterize a detected breast abnormality (Fig. 1). Although the primary care physician is crucial in determining the pattern of subsequent referral, the precise means by which this is achieved varies considerably between different European centres. The management of a particular patient will depend on factors such as age, parity, the persistence of clinical signs or symptoms, the proximity to the menstrual period and the overall anatomy of the breast.

Multiple view, magnification, compression or coned mammograms may be used as an aid to clinical judgement in determining whether calcification or suspicious areas of distortion are present. In women with palpable breast masses techniques such as fine-needle aspiration cytology, which are minimally invasive, may be used in an attempt to establish a diagnosis. Persistent or suspicious palpable masses associated with other anatomical signs or non-palpable masses with suspicious radiographic features, may then be referred for more invasive, surgical evaluation including core-needle or open biopsy [6-9].

The use of BBE in improving diagnosis in women with palpable or non-palpable breast lesions is an area of considerable clinical interest [7]. To be able to deploy, at the point of care, a method which identifies women with both a high and low probability of malignancy would result in considerable savings in time, money and patient anxiety.

Currently, breast lesions which remain of concern despite all efforts at diagnosis short of excisional biopsy pose a particular problem. The present diagnostic process has an overall positive predictive value of less than 60% even in experienced centres. This translates into a benign biopsy rate approaching 30%. In some centres this means that the diagnostic yield of cancers in patients taken to biopsy may be less than 60%. An additional diagnostic test, with a relatively high negative predictive value and an incremental positive predictive value, could therefore be of considerable help in further refining the population of patients subjected to invasive procedures or open surgical biopsy.

The BBE might therefore be helpful at two places within the current process of patient evaluation (Fig. 2). In the screening pathway, BBE could be used in the 5-10% of patients recalled for further evaluation by national mammography programmes. A positive finding would rapidly identify those lesions requiring more expeditious invasive assessment. BBE would also provide a means of following up women with possible abnormalities who did not initially have a pattern of tissue electropotentials indicative of high risk. In this context, the independence of the BBE and radiographic techniques is useful.

Secondly, following referral to hospital (the second pathway in Fig. 2), BBE may constitute a rapid and objective diagnostic technique. The incremental positive predictive value and significant negative predictive value associated with the objective test report could add to the accuracy of clinical decision making and the efficiency of diagnosis. Where initial suspicions have not been resolved by mammography or ultrasound, the results of BBE may support a decision to follow the patient clinically and avoid the unnecessary biopsy of benign lesions or invasive procedures on women unlikely to have breast cancer.

Staging and Selection of Treatment

The BBE may provide an *in vivo* measurement reflective of the underlying proliferative state of the breast epithelium associated with a breast cancer and this may result in an opportunity to employ this novel methodology in treatment planning.

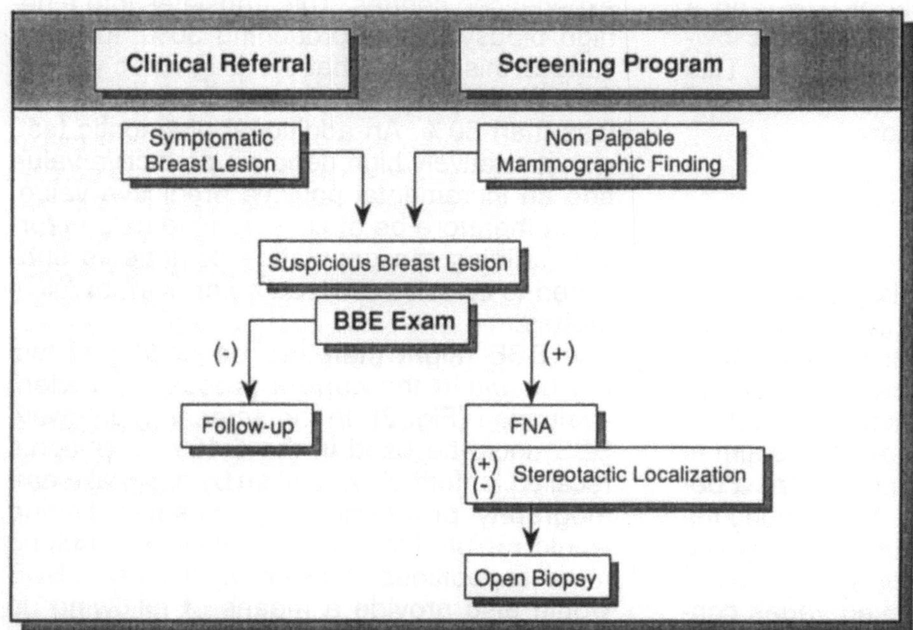

Fig. 2. Measurement of surface electropotentials at various stages of breast cancer management

Independent of tumour size, the biological characteristics of the cancer itself clearly has a role in determining the most appropriate treatment [10-14]. A large but slow growing tumour may be appropriately treated by limited resection of the breast, while a small but aggressive neoplasm may require systemic treatment (chemotherapy and/or hormonal treatment) to prevent distant spread [10]. Given the undoubted influence of differences in proliferative activity on prognosis [10-14], the BBE may provide valuable prognostic information to guide the decision on local or systemic therapy.

One of the major recent advances in breast cancer has been the clinical validation of conservative surgery combined with radiotherapy of non-involved tissue [15,16]. This has resulted in improved cosmetic outcome and the preservation of local lymphatic drainage. However, there is a small but definite rate of local relapse in women who have undergone only limited resection, frequently necessitating a second operation [17-20]. By directly measuring the degree and intensity of cellular proliferation in the area surrounding the tumour, BBE may help in better defining the margins of resection and may also aid identification of women at high risk of local recurrence if treated conservatively. The technique also has the potential to provide a biological marker for patients likely to benefit from adjuvant chemotherapy.

The potential roles of BBE in treatment selection are the following:

- Identification of patients suitable for lumpectomy plus radiation vs more radical surgery.
- Identification of patients requiring more intensive systemic therapy by assessing the rate of tumour proliferation.
- Monitoring effectiveness of cytoreductive systemic therapy.

Detection of Local Recurrence

Radiotherapy results in fibrotic changes which may make the detection of local recurrence or development of a second primary in the previously treated breast difficult [23]. Even when follow-up is strict, current imaging and sampling techniques are associated with a low specificity partly because the sequence of surgery and radiation limits the value of physical examination and mammography. If BBE is associated with characteristic differences in tissue electropotential, earlier recognition of local treatment failure may be possible. Data from Milan involving 14 cases treated with conservative surgery and breast irradiation suggest that BBE is able to detect local recurrence. Clearly, this is an area for future investigation.

Monitoring Chemotherapy

Patients with locally advanced disease or a large tumour burden are treated frequently by preoperative cytoreductive chemotherapy [24, 25]. Recently, chemotherapy has been shown to be successful in reducing local disease to an extent that allows conservative surgery to be undertaken with a reasonable cosmetic result [26]. If BBE can quantify cellular proliferation it may be of use in monitoring response to systemic chemotherapy. A reduced differential value following a cycle of chemotherapy may provide biophysical evidence that cell proliferation is being suppressed. On the other hand, the failure to find such an effect may provide objective evidence of non-responsiveness.

The Future

There is now a strong rationale for the BBE to be studied in populations of women at high risk of breast cancer and particularly in younger women, using a screening array with localization capacity. The increased positive predictive value and negative predictive value of the new test should be evaluated in clinical trials which reflect patient populations undergoing diagnostic evaluation including open surgical biopsy. This will permit objective correlation between BBE and underlying tissue pathology. The potential of BBE to improve the clinical management of populations of patients known to have breast cancer provides an exciting opportunity for clinical research in the next few years.

REFERENCES

1   Brow ML and Fintor L: Cost-effectiveness of breast cancer screening: preliminary results of a systematic review of the literature. Breast Cancer Res Treat 1993 (25):113-118

2   Van Ineveld BM, van Oortmarssen GJ, de Konig HJ, Boer R, van der Maas PJ: How cost-effective is breast cancer screening in different EC countries? Eur J Cancer 1993 (29A):1663-1668

3   Carter R, Glasziou P, van Oortmassen G, de Koning H, Stevenson C, Salked G, Boer R: Cost-effectiveness of mammographic screening in Australia. Aust J Public Health 1993 (17):42-50

4   Miller AB: The cost and benefit of breast cancer screening. Am J Prev Med 1993 (9):175-180

5   Harper GR and Englishbe BH: Prevention and screening for breast cancer. Cancer Detect Prev 1993 (17):551-555

6   Healey EA, Osteen RT, Schnitt SJ, Gelman R, Stomper PC, Connolly JL, Harris JR: Can the clinical and mammographic findings at presentation predict the presence of an extensive intraductal component in early stage breast cancer? Int J Radiat Oncol Biol Phys 1989 (17):1217-1221

7   Randall T: Varied mammogram readings worry researchers. JAMA 1993 (269):2616

8   Lannin DR, Harris RP, Swanson FH, Edwards MS, Swanson MS, Pories WJ: Difficulties in diagnosis of carcinoma of the breast in patients less than fifty years of age. Surg Gyn Obstet 1993 (177):457-462

9   Lindfors KK and Rosenquist CJ: Needle core biopsy guided with mammography: a study of cost-effectiveness. Radiology 1994 (190):217-222

10   Valagussa P, Zambetti M, Bonadonna G, Zucali R, Mezzanotte G, Veronesi U: Prognostic factors in locally advanced noninflammatory breast cancer. Long-term results following primary chemotherapy. Breast Cancer Res Treat 1990 (15):137-147

11   Silvestrini R, Daidone MG, Valagussa P, Salvadori B, Rovini D, Bonadonna G: Cell kinetics as a prognostic marker in locally advanced breast cancer. Cancer Treat Rep 1987 (71):375-379

12   Silvestrini R, Daidone MG, Valagussa G, Di Fronzo G, Mezzamotte G, Mariani L, Bonadonna G: ^{3}H Thymidine labeling index as a prognostic indicator in node positive breast cancer. J Clin Oncol 1990 (8): 1331-1326

13   Meyer JS and Province M: Proliferative index of breast carcinoma by thymidine labeling: prognostic power independent of stage, estrogen and progesterone receptors. Br Cancer Res Treat 1988 (12): 191-204

14   Tubiana M, Pojovie MH, Koscielny S, Chavaudra N, Malaise E: Growth rate, kinetics of tumor cell proliferation and long-term outcome in human breast cancer. Int J Cancer 1989 (44):17-22

15   Fisher B, Redmond C, Poisson R et al: Eight-year results of a randomized clinical trial comparing total mastectomy and lumpectomy with or without irradiation in the treatment of breast cancer. N Engl J Med 1989 (322):822-828

16   Veronesi U, Banfi A, Salvadori B et al: Breast conservation is the treatment of choice in small breast cancer: long-term results of a randomized trial. Eur J Cancer 1990 (26):668-670

17   Holland R, Connolly JL, Gelman R et al: The presence of an extensive intraductal component following a limited excision correlates with prominent residual disease in the remainder of the breast. J Clin Oncol 1990 (8):113-118

18   Boyages J, Recht A, Connolly JL et al: Early breast cancer; predictors of breast recurrence for patients treated with conservative surgery and radiation therapy. Radiother Oncol 1990 (19):29-41

19   Solin LL, Fowble BL, Schultz DJ, Goodman RL: The significance of the pathology margins of the tumor excision on the outcome of patients treated with definitive irradiation for early stage breast cancer. Int J Radiat Oncol Biol Phys 1991 (21):279-287

20   Fowble B, Yeh IT, Schultz DJ et al: The role of mastectomy in patients with stage I-II breast cancer presenting with gross multifocal or multicentric disease of diffuse microcalcifications. Int J Radiat Oncol Biol Phys 1993 (27):567-573

21   Recht A and Harris JR: Selection of patients with early-stage breast cancer for conservative surgery and radiation. Oncology-Huntingt 1990 (4):23-30

22   Galinsky DL, Sharma M, Harsell WF, Griem KL, Murthy A: Primary radiation therapy to T1 and T2 breast cancer following conservative surgery. Which patients should be boosted? Am J Clin Oncol 1994 (17):60-63

23   Recht A, Sadowsky NL, Cady B: Clinical problems in follow-up of patients after conservative surgery and radiotherapy. Surg Clin North Am 1990 (70):1179-1186

24   Hortobagyi GN, Blumenschein GR, Spanos W, Montague ED, Buzdar AU, Yap IIY, Schell F: Multimodal treatment of locoregionally advanced breast cancer. Cancer 1983 (51):763-768

25   Feldman LD, Hortobagyi GN, Buzdar AU, Ames FC, Blumenschein GR: Pathological assessment of response to induction chemotherapy in breast cancer. Cancer Res 1986 (46): 2578-2581

26   Bonadonna G, Veronesi U, Brambilla C et al: Primary chemotherapy to avoid mastectomy in tumors with a diameter of three centimeters or more. JNCI 1990 (82): 1539-1545

Use of Non-Directed (Screening) Arrays in the Evaluation of Symptomatic and Asymptomatic Breast Patients

Joseph P. Crowe, Jr. [1] and Mark L. Faupel [2]

1   Director, Breast Services, The Cleveland Clinic Foundation, 9500 Euclid Avenue, Cleveland, Ohio 44195, U.S.A.
2   The University of Georgia, Boyd Graduate Studies, Research Center, Brooks Drive, Athens, GA 30602, U.S.A.

Early detection of breast cancer by screening is the only means for long-term survival. In the United States, breast cancer screening recommendations for asymptomatic women include breast self-examination every month beginning at age 20, clinical breast examination every 3 years between ages 20-40 and every year thereafter, and mammography every 1-2 years between ages 40-49 and every year thereafter [1]. Results of several large breast cancer screening studies over the past 30 years have shown up to a 30% relative reduction in mortality but less than a 1% absolute reduction [2]. Whether efforts to increase breast cancer screening during the past decade will reduce mortality should be apparent over the next several years.

Ideally, a screening test is easily administered, non-invasive, repeatable, inexpensive, and gives an objective result. A cancer screening test should be able to identify disease prior to systemic dissemination and allow for appropriate interventions to reverse the process, or at least arrest its continued development. As discussed in previous chapters, the observation that upregulation of epithelia can be identified by electrical depolarizations, even prior to morphologic change, may have important implications for cancer screening. For breast cancer, the possibility that electrical depolarizations can be detected using skin surface measurements leads to intriguing possibilities for screening. Previous studies have examined whether breast skin surface potentials provide diagnostic information for previously identified suspicious breast lesions [3]. These studies utilized an array of sensors placed on an individual quadrant of the breast and have shown that this technology can discriminate benign from malignant breast lesions with high sensitivity and specificity.

Although it appears that the skin surface electropotentials can be used for diagnosis, placement of the sensor array depends upon identification and localization of a lesion by mammography or physical examination (palpation). Such an array could not be used for screening of asymptomatic patients who do not have localized abnormalities. The first question in evaluating this technology for screening is whether a non-directed array of sensors which does not depend upon a knowledge of location of the lesion, can be used. To evaluate this, patients who were going to breast biopsy were studied using a non-directed array of sensors which did not depend upon the location of the lesion. The primary aim of this study was to evaluate whether an array of sensors which covered all four quadrants in both breasts in a mirror-image fashion, would yield results similar to the previously used diagnostic array. This so-called non-directed, or screening, array consisted of placing 16 sensors on each breast, a sensor in each axilla, and a reference sensor on each palm. The sensors were placed in the same position for each patient. The sensitivity and specificity of this array for discriminating cancer from benign lesions was compared to that of the directed or diagnostic array for 3 groups of patients: those with biopsy-proven cancer, those with biopsy-proven benign lesions, and those without breast symptoms who did not go to biopsy and who were considered to be normal. This was a feasibility study and represented a first attempt to evaluate sensor arrays which might eventually be adapted for use in breast screening.

Methods

Patient Population

One hundred and twenty-seven women were enrolled over a 6-month study conducted at Case Western Reserve University, Cleveland, Ohio, U.S.A. To be eligible, patients had to be at least 18 years of age and provide informed consent. Patients were excluded if they had a prior breast cancer history or breast implants.

Procedures

All patients had breast surface electrical potentials measured using a 38-channel electropotential scanning device manufactured by Biofield Corp. of Roswell, Georgia, U.S.A. After informed consent was obtained, skin surface sensors were attached to both breasts in a standard pattern. Sixteen sensors were placed on each breast in two concentric circles; an inner circle of 4 sensors was placed just outside the areolae (at the 1:30, 4:30, 7:30, and 10:30 positions), and an outer ring of 12 sensors at each of the 12-hour clock positions. Two additional sensors were placed under the axillae and reference sensors were placed on each

palm. Figure 1 provides a graphic representation of the sensor placement used for all patients in the study.

After all sensors were positioned and patient medical history recorded, measurement of skin potentials began. For each sensor in the array, 150 measurements were averaged to produce a composite voltage relative to the palm sensors, for each point on the breasts and axillae. Sequencing, timing and averaging of the 150 individual surface electropotential measurements was done automatically by computer using custom software. The measurement phase of the study was about 4.5 minutes.

Data Analysis

Analysis of surface potentials focussed on variables previously identified as providing information which could be used to discriminate between benign and malignant conditions. These measurements included the Within-Breast Differential ("WBD"), which is the difference between the highest and lowest voltages on the involved breast, the Between-Breast Differential ("BBD"), which is the difference in the WBDs between the two breasts, and the absolute lowest averaged potential from the involved breast ("LOW"). An additional new measure of between-breast disparity also was evaluated. This is the Mirror Site Differential ("MSD"), which is the average difference between all 19 pairs of corresponding sensors. It can be considered as a more precise measure

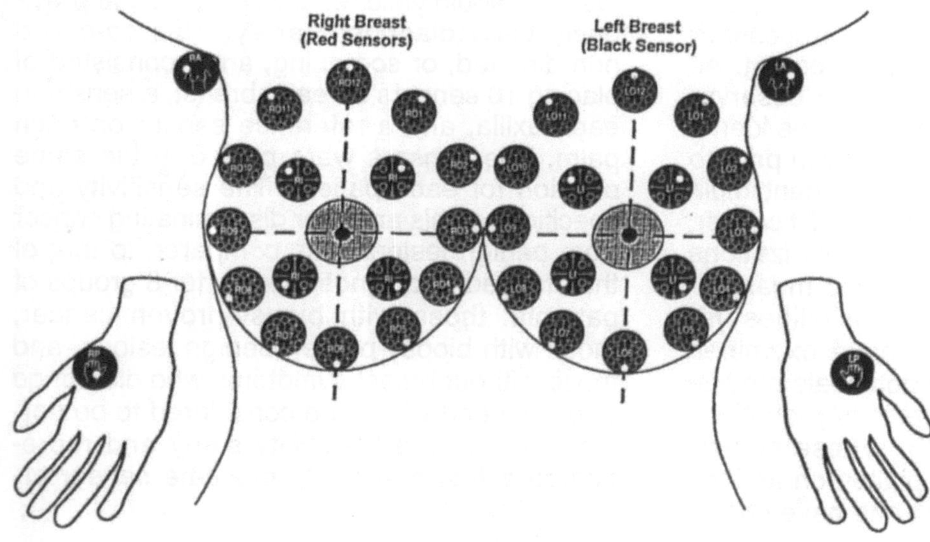

Fig. 1. Non-directed (screening) sensor array

of surface potential asymmetry between two breasts since it takes into account the spatial arrangement of the sensor arrays for each patient.

Differentials for the four measures described above were compared for the different disease categories. Previous studies [3] have indicated that surface potential data within disease categories are not distributed normally. Therefore, non-parametric statistical methods were used to both demonstrate differences between groups and to formulate decision-making models based on surface potential differences. In the former case, the Kruskal-Wallis One Way ANOVA on Ranks was used to summarize differences between 3 or more groups, while the Mann-Whitney U Test was appropriate for 2-grouping comparisons. In the latter case, non-linear decision trees were developed using dedicated software such as CART (Classification and Regression Trees).

Results

The 127 patients enrolled in the study ranged in age from 18 to 93 years and were about evenly distributed above and below 50 years of age (see Table 1). Ninety-five of the 127 patients had breast lesions which were biopsied after surface electrical potential measurements were taken. The remaining 32 patients were volunteers who were tested but did not proceed with biopsy because they did not present with a localized breast lesion.

After surface potential measurements, all 95 patients with breast lesions had excisional biopsies performed for histological diagnosis. The remaining 32 patients did not go to biopsy because they were asymptomatic and did not have breast lesions identified. Three of the biopsy patients had bilateral lesions which could be analyzed separately resulting in a total of 98 biopsied lesions included in the study. Of the 98 lesions biopsied, 51 (52%) were palpable and 47 (48%) were non-palpable and were identified mammographically. Clinical evaluation of both palpable and non-palpable lesions revealed a size range from 0.5 cm to 6.0 cm for cancers and 0.5 cm to 7.0 cm for benign lesions.

Of the 98 biopsied lesions, there was a range of benign, pre-malignant and malignant lesions.

Table 1. Age distribution of patients tested

Disease state	No. pts. < 50 yrs	No. pts > 50 yrs	Age range (years)
Cancer (n=27)	8	19	32-93
Benign (n=68)	27	41	18-82
Asymptomatic (n=32)	27	5	24-60
All patients	**62**	**65**	**18-93**

Table 2. Histological results of biopsied lesions

	Number of lesions
Invasive ductal or lobular carcinoma	21
Ductal carcinoma *in situ*	7
Lobular carcinoma *in situ*	1
Atypical ductal or lobular hyperplasia	16
Papillomas	5
Fibrocystic changes	29
Fibroadenomas	9
Other / Non-proliferative	5

There were 28 malignant lesions, of which 21 were invasive ductal or lobular carcinoma and 7 were non-invasive ductal carcinoma *in situ*. Included among the 69 benign lesions, there were both proliferative and non-proliferative lesions. Table 2 gives the frequency distribution of histological diagnoses.

Electropotential measurements varied as a function of disease state. The average values of 4 of these readings (the absolute low reading, within-breast differential, between-breast differential, and mirror site differential) differed significantly among the three disease states studied (cancer, benign, asymptomatic). The differences were especially pronounced with regard to the whole breast differential of the symptomatic breast, the absolute lowest potential of the symptomatic breast and the mirror site differential (see Table 3). Cancers produced the highest levels of both absolute depolarization (absolute low reading) and relative depolarization (within-breast differential, be-

Table 3. Mean key electropotential measures (in mV) as functions of disease state

Disease state	Absolute low reading	Within-breast differential	Between-breast differential	Mirror site differential
Cancer (n=27)	0.6	39.1	18.0	10.3
Benign (n=68)	15.0	26.3	12.9	7.2
Asymptomatic (n=32)	13.7	21.8	7.6	6.1
P =	0.001	0.002	NS	0.002

tween-breast differential, and mirror site differential) (see Table 4).

The variation in differentials for cancer, hyperplasias, benign, and unbiopsied disease states can be shown graphically in 2-dimensional plots. The between-breast and within-breast differentials, with standard errors of the means, can be plotted on the Y and X axes, respectively, for the 28 malignant lesions and 70 benign lesions as shown in Figure 2. Using a similar graph, the 70 benign lesions were separated into 43 non-hyperplastic and 27 hyperplastic lesions and the distributions for the respective differentials for cancer, hyperplasia, non-hyperplasia and unbiopsied patients are shown in Figure 3.

Averaged potential measurements and differentials were inserted into the CART programme in order to produce a decision tree which could be used retrospectively to generate sensitivity and specificity for the biopsied patients. The decision tree produced by CART was limited to 10 decision points, or nodes, and produced a sensitivity of 100% (28/28) and a specificity of 87% (61/70). Since the sample size was small, there were no clear trends regarding the types of benign lesions classified as false positives.

Table 4. Symptomatic within-breast differentials (WBD) as a function of disease state

Disease state	Number studied	Mean WBD (mV)
Cancer		
Invasive	21	41.3
Ductal carcinoma *in situ*	7	32.4
Hyperplasia (proliferative)		
Atypical hyperplasia/LCIS*	6	29.0
Hyperplasia/Papilloma	21	28.9
Benign (non-proliferative)	43	24.7
Unbiopsied		
Asymptomatic volunteers	32	21.8

* Lobular carcinoma *in situ*

Discussion

The objective of this study was to compare results of a non-directed (or screening) array of sensors with results of the directed (or diagnostic) array of sensors. Absolute and/or relative depolarizations reflected in the absolute low reading and the within-breast differential, between-breast differential, and mirror site differential, respectively, were found using the non-directed sensor arrays. Two variable differential comparisons show clearly that malignant lesions can be separated from benign lesions (see Fig. 3). These results are similar to those for the diagnostic array. Furthermore, using the CART algorithm a 100% sensitivity and an 87% specificity were obtained using the non-directed arrays. These results were independent of patient age and lesion size, with lesions as small as 0.5 cm producing abnormal levels of depolarization.

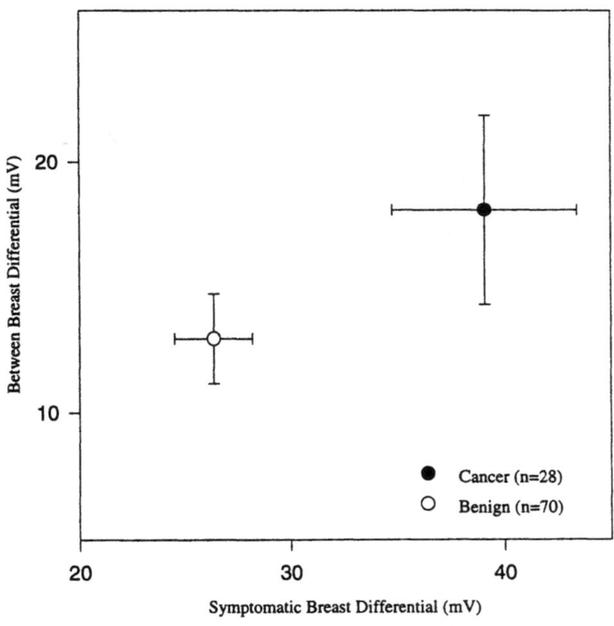

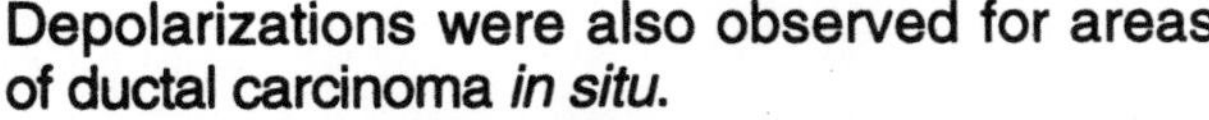

Fig. 2. Mean electropotential differentials (in mV) with standard error bars for malignant and benign lesions

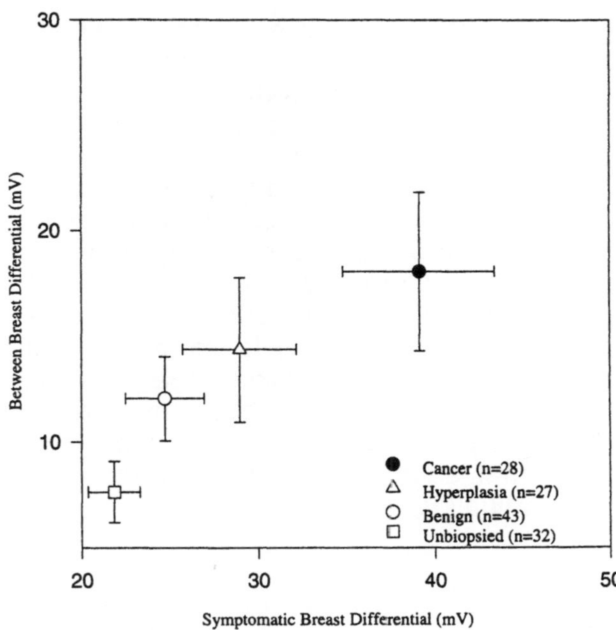

Fig. 3. Mean electropotential differentials (in mV) with standard error bars by disease state

Depolarizations were also observed for areas of ductal carcinoma *in situ*.

There were definite differences in depolarization and thus differentials comparing the different arrays as shown in Tables 2 and 3 and Figure 3. The magnitude of the differentials was most striking for the cancer group, as would be anticipated from previous studies using directed sensor arrays [3].

These data certainly suggest that it is not necessary to use a directed array of sensors to accurately discriminate among patients who have benign and malignant lesions identified by palpation and/or mammography. The Biofield Test would certainly appear to fulfil the criteria for a potential screening application: it is easily administered, non-invasive, repeatable, inexpensive, and yields an objective result. Application of such a test for screening purposes raises several issues. From studies of colon epithelia, depolarizations of upregulated regions have been shown to precede histological changes [4]. This would suggest that, if this test were used as an initial screening modality, patients could potentially be identified who are in early stages of cancer development and who do not have a localized lesion identified by either mammography or palpation. If such patients without identifiable lesions were found to have regions of depolarization, the test could

be considered a "false positive". However, these patients could have upregulated proliferating epithelium without histological, mammographic, or palpable findings, and be at increased risk for developing breast cancer. Correlation of histological changes with skin surface depolarization could be determined by obtaining breast biopsy samples from depolarized regions. Localization of depolarized regions is theoretically possible using vector interpolation to estimate surface potential levels between each of the measuring sensors. Assuming the same number of sensors were used for all patients, variation in breast size and contour would result in different distances between sensors which would need to be considered in the vector model. If depolarization regions were not biopsied, patients would need to be followed for development of a mammographic or palpable lesion.

The truly intriguing aspect of this important and novel biophysical evaluation is that it would yield a different type of measurement, one not relying upon the development of an abnormal mammogram or breast mass. Perhaps identification of patients who are in a phase of breast cancer development during the 10 years prior to either mammogram or physical exam changes would allow for a significant impact on breast cancer outcome. Perhaps such a test would serve to identify a group of patients who are at risk for development of systemic breast cancer who would then be candidates for intervention studies to possibly reverse the biophysical measurements.

To question whether measurements of skin surface depolarizations could be used for breast cancer screening in an asymptomatic population, and ask how these results should be compared to screening efforts using palpation and mammography, is certainly valid. However, skin surface potential measurements offer such a unique and novel means of identifying both possible high-risk and cancer patients, that such a modality deserves very close evaluation.

REFERENCES

1   Mettlin C and Smart CR: Breast cancer detection guidelines for women aged 40 to 49 years: Rationale for the American Cancer Society reaffirmation of recommendations. CA - A Cancer Journal for Clinicians 1994 (44):249

2   Wright CJ: Breast cancer screening: A different look at the evidence. Surgery 1986 (100): 594-598

3   Weiss BA, Ganepola GAP, Freeman HP, Hsu Y-S, Faupel ML: Surface electrical potentials as a new modality in the diagnosis of breast lesions - a preliminary report. Breast Disease 1994 (7):91-98

4   Goller DA, Weidema WF, Davies RJ: Transmural electrical potential difference as an early marker in colon cancer. Archives of Surgery 1986 (121):345-350

The Value of BBE in the Assessment of Breast Lesions

Martina Dickhaut [1], Ingrid Schreer [1], Hans-Joachim Frischbier [1] (German data) and
Mirella Merson [2] (Italian data)

1   Department of Gynaecological Radiology, University Hospital of Hamburg, Martinistrasse 52, 20246 Hamburg,
    Germany
2   Department of Surgical Oncology, Istituto Nazionale Tumori, via Venezian 1, 20133 Milano, Italy

This chapter describes the initial clinical experiences with Breast Biophysical Examination (BBE) in two European countries: Germany and Italy. The primary aim of these feasibility studies was the transfer of this technology to novel environments. As with any new technology, care must be taken when transferring it to new environments. In many cases, the technology must be adapted to the local situation in order to realize its full potential.

The plan for validating BBE in Europe has two phases. In the initial phase, all factors which may impact on the use of the technology must be carefully evaluated. These include assessment of clinical need, technician training, training validation, education of clinicians, modification of system hardware and software (if required), and preparation for the second phase of validation, which is the execution of a blinded, multicentre clinical trial in Europe.

During the initial phase of technology validation, data are collected and analysed in an open, collegial manner. If study protocols for this phase do not include masking provisions, it is important that data reviews occur on a regular basis. Study sponsors, scientists and principle investigators must be encouraged to communicate freely. (One reason for the formation of the European School of Oncology task force to study BBE was to foster this type of international communication.) The studies described in this chapter summarize the culmination of this initial phase of clinical validation.

Methods

BBE testing occurred at two clinics in Germany (Gynaecological Clinic of Hamburg University and Esslingen Hospital), as well as at the Istituto Nazionale per lo Studio e la Cura dei Tumori (INT) in Milan. The methodology employed for evaluation of BBE was similar for both the Italian and German populations. The same prototype device was used for testing all patients enrolled in the study. The device and procedures used for testing patients have been described in detail by Faupel and Hsu in this volume and will be summarized only briefly here. The procedure was explained to each patient and informed consent was obtained prior to testing. Patients were included in the study if they had a previously localized palpable or non-palpable breast lesion scheduled for open biopsy. Patients were excluded from analysis if they had undergone preoperative chemotherapy or were taking tamoxifen. Once the suspicious lesion was located, one sensor was placed directly over the centre of the lesion and 4 additional sensors were placed around the margins of the lesion in a north-south, east-west pattern. Two additional sensors were placed in quadrants adjacent to the involved quadrant, and one sensor was placed under the axilla. The reference or ground sensor was placed on the ipsilateral thenar eminence (palm). The minimal allowable spacing between sensors was 3.3 cm from centre to centre of the sensors. This pattern of sensor placement was then reproduced in a mirror-image fashion on the contralateral breast, axilla, and palm.

Once all sensors were positioned securely and case report forms completed, recording of breast surface potentials relative to each of the palm references began. Sequencing and timing of electropotential (EP) sampling was under computer control. EP recording lasted approximately 2.5 minutes and consisted of 150 potential measurements taken from each sensor on the breasts and axillae, first sampled relative to the left palm reference and then resampled relative to the right palm reference. Once all measurements were taken, the computer averaged the 150 potential samples from each reference to produce a composite voltage for each sensor.

Data analysis was similar to that previously developed for the preliminary US studies. These analyses included calculation of EP differentials, both within the involved breast and between the two breasts. In the original US studies (which employed a previous device prototype) an array of 6 sensors was used on each breast and axilla: 5 sensors on the breast and one on the axilla. In Europe, it became apparent that consistent placement of the axillary sensors was not possible due to interference from axillary hair. The prototype device used in Europe permitted the use of 4 additional recording channels which allowed EP measurements to be taken from 2 additional quadrants from each breast. Therefore, test data from European clinics are summarized using differentials calculated from the 7 sensors placed on each breast and do not include the axillary EP readings.

German Data

Data from April 1993 until the beginning of August 1994 were analyzed by the study sponsor. During this testing period, two different types of sensors were used, the more effective of which will be used in the next phase, a double-blind multicentre study. The results summarized in this chapter include tests using the sensor which will be employed in the double-blind multicentre study.

Results from Hamburg and Esslingen were pooled. The test population for this study consisted of 158 patients. Sixty-four were tested in Hamburg and 94 in Esslingen. The age distribution is given in Tables 1 and 2.

The percentage of non-palpable lesions in Hamburg was 42.3% and in Esslingen 30.1%. In both groups tested, the majority were palpable lesions, 57.8% of the tested patients in Hamburg and 68% in Esslingen. Clinical sizes of the tumours tested are summarized in Tables 3 and 4.

Of the 64 patients tested in Hamburg, 65.6% had malignant lesions and 34.4% benign lesions. In Esslingen, 47.9% of the 94 had malignant lesions and 52.1% had benign lesions. The malignant lesions included invasive as

Table 1. Hamburg University Hospital - Age distribution of patients

Disease state	Age range	50 years or less	Age over 50
Cancer	36-85	12	30
Benign	32-76	13	9
All patients	32-85	25	39

Table 2. Esslingen Hospital - Age distribution of patients

Disease state	Age range	50 years or less	Age over 50
Cancer	42-86	6	39
Benign	18-80	20	29
All patients	18-80	26	68

Table 3. Hamburg University Hospital - Clinical size evaluation of lesions tested

Disease state	Non-palpable lesions	Palpable lesions	Size range of palpable lesions
Cancer	14	28	1.0-5.0
Benign	13	9	1.0-4.0
All lesions	27	37	1.0-5.0

Table 4. Esslingen Hospital - Clinical size evaluation of lesions tested

Disease state	Non-palpable lesions	Palpable lesions	Size range of palpable lesions
Cancer	9	35	1.0-15.0
Benign	20	29	1.5-4.0
All lesions	29	64	0.5-15.0

The size of one malignant tumour was not reported

well as non-invasive carcinomas. The benign lesions showed a variety of histopathological appearances, including atypical hyperplasia, adenosis, fibroadenoma, fibrocystic changes, hyperplasia, intraductal papilloma, and lobular carcinoma *in situ*. The distribution of histopathological findings is given in Tables 5 and 6.

The sensitivity and specificity of the test results were determined retrospectively using CART (Classification and Regression Tree) software.

CART analysis of EP differentials produced 2 false-negative results, both in patients with invasive ductal carcinomas. Three false-positive results were identified, one of which was atypical hyperplasia, one a radial scar, and the third simply lipomatous parenchyma. Our test results showed a sensitivity of 95.2%, a specificity of 86.4%, a positive predictive value of 93% and a negative predictive value of 90.5%. In the Esslingen test population a sensitivity of 93.3%, a specificity of 73.5%, a positive predictive value of 76.4% and a negative predictive value of 92.3% were seen.

Table 5. Hamburg University Hospital - Histopathology results

Disease state		Number of lesions studied
Cancer		
	Invasive cancer	37
	Ductal carcinoma *in situ*	5
Benign		
	Atypical hyperplasia	1
	Adenosis	0
	Fibroadenoma	5
	Fibrocystic changes	5
	Hyperplasia	0
	Intraductal papilloma	1
	Lobular carcinoma *in situ*	0
	Apocrine metaplasia	0
	Miscellaneous benign	10

Table 6. Esslingen Hospital - Histopathology results

Disease state		Number of lesions studied
Cancer		
	Invasive cancer	40
	Ductal carcinoma *in situ*	5
Benign		
	Atypical hyperplasia	0
	Adenosis	17
	Fibroadenoma	4
	Fibrocystic changes	10
	Hyperplasia	5
	Intraductal papilloma	4
	Lobular carcinoma *in situ*	1
	Apocrine metaplasia	1
	Miscellaneous benign	7

Table 7. Hamburg and Esslingen - Results of retrospectively analyzed data

Site	Sensitivity	Specificity	PPV	NPV
Hamburg	95.2% (40/42)	86.4% (19/22)	93.0% (40/43)	90.5% (19/21)
Esslingen	93.3% (42/45)	73.5% (36/49)	76.4% (42/45)	92.3% (36/39)
Total	94.3% (82/87)	77.5% (55/71)	83.7% (82/98)	91.7% (55/60)

PPV = positive predictive value; NPV = negative predictive value

Discussion

The Breast Biophysical Examination is a new non-invasive technique which can be performed easily by a physician or a technician. It is not associated with any significant risk to the patient's health. We did not see any adverse effects, not even an allergic reaction to the contact gel/cream. The testing was well accepted by the patients and none of the women tested complained of any discomfort. The whole procedure, including examination of the breasts, placing of the sensors and completion of the case report forms, took approximately 20-25 minutes per patient.

It would be useful to have an additional diagnostic method for breast lesions of sufficiently high sensitivity and specificity which complements mammography, ultrasound and MRI in order to differentiate between benign and malignant lesions. Although the preliminary results described in this chapter were analyzed retrospectively, they are encouraging. In a double-blind study it will be proven whether the BBE objectively produces such good results, and whether it will be able to fill the diagnostic gap that currently exists.

Italian Data

As an important national and European centre for cancer research and therapy, the Istituto Nazionale per lo Studio e la Cura dei Tumori of Milan (INT) was able to enrol large numbers of patients in the clinical assessment of this new technology. Several surgical departments collaborated in the validation testing, which allowed rapid training of several technicians and standardization of methods and materials necessary for clinical evaluation of BBE. The study protocol involved testing women with previously localized breast lesions who were scheduled for biopsy and/or surgical treatment. The initial study involved validating the performance of a new prototype BBE device, testing the same patient twice, once with the M2 BBE device and once with the newer M3 device. This allowed us to determine whether the M3 device was as accurate as the M2 device, which had demonstrated efficacy in previous clinical studies conducted in the US.

Additional protocols have focussed on evaluating different types of sensor systems, including a within-subject comparison of two different sensor arrays. These studies are currently in the analysis phase. The latest version of the device (the M4) is now in service at INT. The M4 version can no longer be considered as a prototype; in design and function it is the device which will be available for clinical use once the ongoing multicentre trials are completed.

The preliminary work at INT involved training and transfer of technology and resulted in improvements in both the device and sensor system. The clinical data described in this section summarize results using a BBE system comparable to the one anticipated for use in the European multicentre blinded study. This multicentre study, which will involve at least 500 women scheduled for open biopsy in 8 leading centres in Europe, will provide a blinded assessment of the clinical effectiveness of this new, non-invasive test. After 16 months of experience comprising hundreds of BBE tests performed at INT, we can attest to the fact that

Table 8. Patient characteristics

Age of patients: 43 % ≤50 yrs	Range: 22 - 87 yrs	
Pathological size of malignant lesions: 25.4% ≤ 1 cm	Range: < 0.2 - 8 cm	
Non-palpable lesions: 23 %	36/191 cases	
Pathological results:	116	malignant
	75	benign
116 malignant:	108	invasive carcinoma
	8	ductal carcinoma *in situ*
75 benign:	including 6 cases of atypical hyperplasia and 1 LCIS	

the device poses negligible risk to the patient, and that patient acceptance of the test has been excellent.

Materials and Methods

The group studied consisted of 191 lesions, which were biopsied after BBE. Nine patients were tested but have not been included in the analysis: 2 because they had implanted electrically conductive material (one infusion pump and one prosthesis valve), 5 because of variable values outside the permitted range, one because of a previous carbon injection and one because of a needle localization wire in the tested breast.

Results and Discussion

Patient characteristics are summarized in Table 8.
The analysis of the BBE results demonstrated significant differences in the differential value between benign and malignant lesions in symptomatic (p = 0.002) as well as in the inter-breast mean differentials (p = 0.007) (Fig.1).
In this group there were 3 false negatives and 14 false positives, resulting in a sensitivity of 97.4% and a specificity of 81.3% for breast malignancy. These first results obtained from the Milan validation studies replicated the preliminary US results.

The 6 atypical hyperplasia cases and the single case of LCIS had a higher differential than other benign lesions. In Table 9 the histological diagnosis has been related to the mean differential: invasive cancers produced the highest differentials (22.2 mV) while non-invasive cancers were somewhat lower (18.9); this difference was not statistically significant. The highest differentials from the benign lesions group were in the hyperplasic group (19.5 mV). Fibroadenomas also produced relatively high

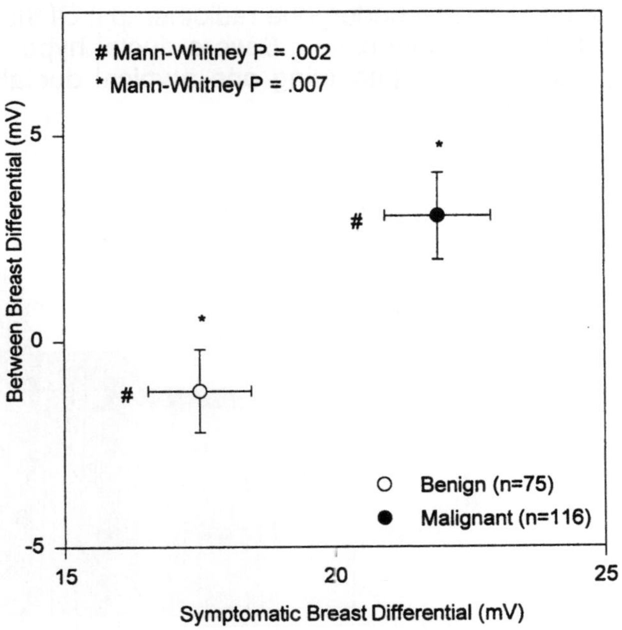

Fig. 1. Mean differentials (with standard error bars) for benign and malignant lesions

Table 9. Mean differential by histopathology result

Disease state	Number of lesions studied	Mean differential
Cancer		
Invasive Cancer	108	22.2
Ductal carcinoma *in situ*	8	18.9
Benign		
Atypical hyperplasia (6), Lobular carcinoma *in situ* (1), Intraductal papilloma (6), Hyperplasia (5)	18	19.5
Fibroadenoma	18	19.5
Adenosis	7	18.7
Apocrine metaplasia	7	16.1
Fibrocystic changes	9	15.5
All other benign	16	14.2

differentials (19.5 mV). The lowest differentials were associated with fibrocystic changes (15.5 mV) and miscellaneous benign lesions (14.2 mV).

The 3 false negatives were a bicentric invasive ductal carcinoma and 2 invasive lobular carcinomas. None of the false negatives involved patients who were currently undergoing or had recently undergone radiotherapy. Of the 14 false-positive cases, 6 were ductal hyperplastic lesions (including one atypical ductal hyperplasia), 5 were fibroadenomas, one was a chronic mastitis associated with fibrosclerosis and one was duct ectasia. Of these 17 diagnostic errors, 5 involved women under the age of 50 and 12 involved women aged 50 or older. The group studied included 12 patients previously treated by breast irradiation after conservative surgery; they were correctly diagnosed by BBE and included 2 non-palpable lesions. This means that previous irradiation does not create an obstacle to surface differential recording and BBE diagnosis.

An additional evaluation in Milan was the correlation between the cellular proliferation measure (Thymidine Labelling Index) and amplitude of mean differential. The analysis included 84 cancers and suggested a relationship between this proliferative parameter and the mean differential recorded from symptomatic breasts. In Figure 2 labelling index (LI) has been divided into 4 categories, corresponding roughly to low proliferation cancers (LI ≤ 0.7), moderate-low proliferation cancers (LI = 0.8 - 2.2), moderate-high proliferation cancers (LI = 2.3 - 4.5), and high proliferation cancers (LI > 4.5). Figure 2 indicates that the mean electropotential differential calculated from the symptomatic breast increases as labelling index increases. The r-squared value calculated for this relationship is 0.51 (p < 0.001). The relationship demonstrates a correlation between tissue neoplastic proliferation and surface potential recordings.

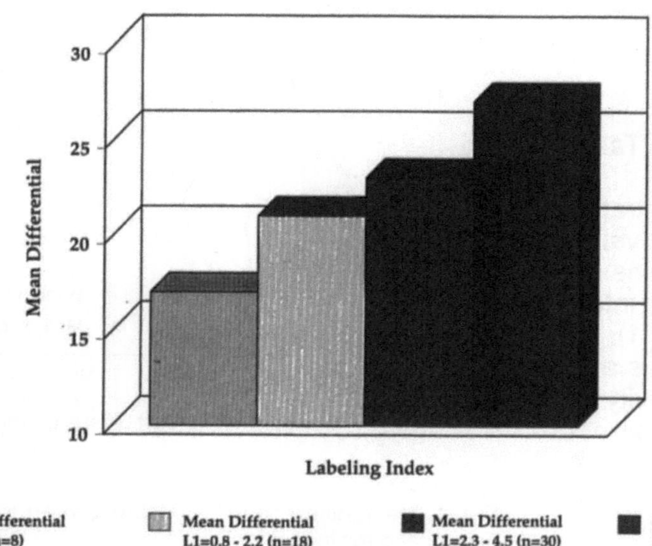

Fig. 2. Relationship between mean differential (symptomatic breast) and labelling index

Comment

This correlation between proliferative activity of neoplastic tissue and the mean differential may explain some of the false positives. If this biophysical evaluation of tissue activity reflects tissue proliferation, it is easy to understand that hyperplastic tissue may be more proliferative than some invasive cancers.

Acknowledgements

Dr. Merson is grateful to the following colleagues for active collaboration in patient testing: Dr. Bernardo Bonanni, Dr. Monica Terenziani, Dr. Ruj Gin, Dr. Luigi Mascheroni and Dr. Catia Bonfatti. Special acknowledgement is given to the silent but continuous data-managing and secretarial activity of Ms Francesca Falcetta.

Multicentre Clinical Trials Evaluating the Breast Biophysical Examination in the Assessment of Breast Disease

Bruce A. Bach

Assistant Clinical Professor of Medicine, University of California, 505 Parnassus Avenue, San Francisco, CA 94143, U.S.A.

The Breast Biophysical Examination (BBE) is meant to be an adjunctive modality, designed to aid in the identification of patients likely to have either a benign or malignant breast lesion. The test is intended to be complementary to mammography and physical examination. The nature and technical requirements of the test are such that its use is of particular potential value relatively early in the sequence of diagnostic evaluation of women with breast lesions of unknown significance.

When assessing the performance of any new diagnostic modality or test, it is important to define at what point in the diagnostic workup the new test is intended to be applied, and to identify clearly the criteria being used to demonstrate the validity of the test result [1-6]. Equally important in assessing utility is the spectrum of disease represented by the patients included in the study [7-9]. Acknowledgement of the necessary limitation imposed by the spectrum of patients and by the characteristics of the chosen independent measure of disease is the key to developing an understanding of the diagnostic discrimination provided by any novel test or technique. This chapter will review the clinical trial study design and quality assurance aspects of ongoing multicentre clinical studies involving the diagnostic application of the BBE.

Work to Date: Basis for a Well Designed Clinical Study

The evaluation of a new diagnostic test begins with the measurement of a representative sample of subjects from both reference and diseased populations [9]. In the case of the BBE, in the pilot trial phase each patient enrolled had been scheduled for open breast biopsy as part of normal clinical decision-making for women at risk of breast cancer at that particular centre. No attempt was made to influence the selection of patients by demographic characteristic or disease stage.

Much of the pilot clinical data obtained to date concerning the operating characteristics of the BBE have been described in the preceding chapters of this monograph. The results suggest that patterns of surface electropotential differentials are demonstrably different in patients with benign versus malignant disease. Furthermore, the available evidence supports the idea that the BBE can provide a reasonable basis for forming a judgement concerning the presence of either benign or malignant breast lesions prior to open surgical biopsy.

The pilot clinical study results reported in this monograph employ two successive generations of measuring instruments and a related family of electroconductive media and sensor platforms which have been used to generate a set of observations for decision-making and analysis.

Using the results of the pilot phase clinical trials, the clinical investigators have established the range of key test parameters and have made a formal analysis of where the informative cutoff values of the measured variables are likely to be based on a sample of the population of interest [10,11]. The second phase involves a prospective blinded study of the BBE's discriminant power in a new trial population unrelated to the first phase patients [2,3, 5].

Elements of Trial Design

In any clinical trial of a new or improved diagnostic test, the nature of the test and the nature of the reported result influence the choice of trial design. In the diagnostic application of the BBE, it is anticipated that the reported result would be an objective statement that a woman has a certain pattern of electropotential differentials associated with the presence of a malignancy, or alternatively that the observed pattern was not associated with malignancy in clinical trial patients.

Additionally, a lesion scoring system measuring the magnitude of serially measured electropotentials might be a clinically desirable output from the BBE, providing the possibility of evaluating temporal trends.

The key to implementing an informative clinical trial of the BBE is a clear definition of where and when in the diagnostic workup the test is intended for use. As shown in Figure 1, patients may enter the diagnostic process from one of two pathways. Asymptomatic patients from an at-risk population may be identified as having a breast lesion after breast cancer screening mammography or other imaging modality. Actual post-mammography screening recall rates vary across Europe, with reported rates ranging between 5-7% of all screening examinations having radiographic features that warrant additional imaging or invasive studies [11-16].

The additional studies, triggered by image-detected lesions, may lead to the identification of a number of breast cancer cases which are eventually verified by tissue examination. However, the specificity of mammographic screening remains relatively low. The availability of the BBE in this setting, where physical examination and first-line imaging results are used for decision-making, would be of significant potential value. In this regard, it is relevant that the multicentre validation trial will include a population of patients who are asymptomatic and who have come to clinical attention solely on the basis of radiographic findings. In practice, the majority of patients with mammographically detected lesions will have non-palpable breast lesions.

Conversely, symptomatic women often have palpable dominant masses. This second group of patients is also of interest. Women with palpable breast abnormalities who are being evaluated for possible breast cancer may have a different post-detection workup. However, both of the above groups of patients, either because of the screening, imaging or the initial examination, will have an identified anatomic quadrant of the breast to be evaluated using the BBE.

In the diagnostic clinical trial, the BBE results will be evaluated in patients with previously localized suspicious or clinically indeterminant breast lesions in the context of all available clinical information after a decision has been made to proceed to open surgical biopsy.

Clinical Evaluation of Possible Breast Cancer

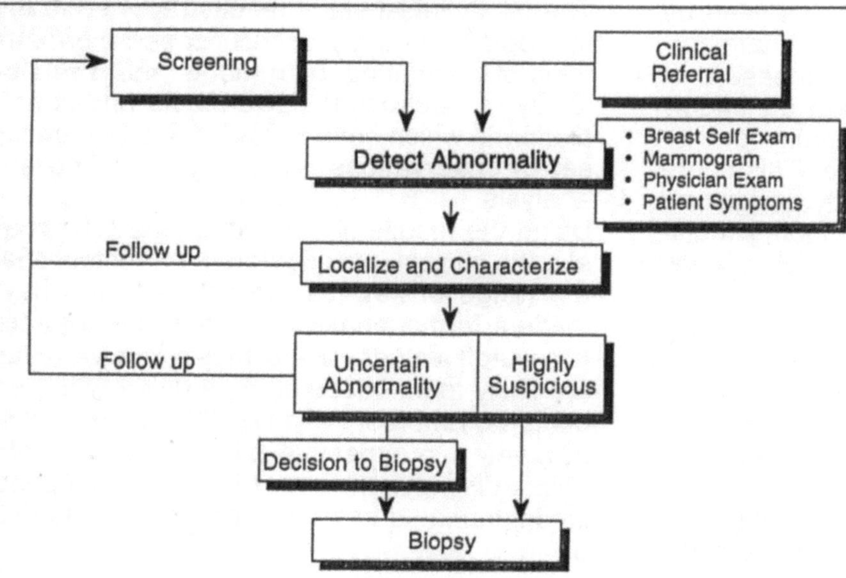

Fig. 1. This diagram defines the stages and key processes in the screening and diagnosis of breast cancer. Women may enter the diagnostic phase after a breast lesion of unknown significance has been detected. The techniques and sequence of actions in the localization and characterization phase show significant variation between centres. After the localization and characterization phase, patients either have highly suspicious findings or remain with uncertain abnormalities. Open biopsy can either reduce uncertainty or confirm a suspected malignancy. Rates of surgical biopsy and yield show variation throughout Europe.

Study Design: Self Controlled Parallel Study

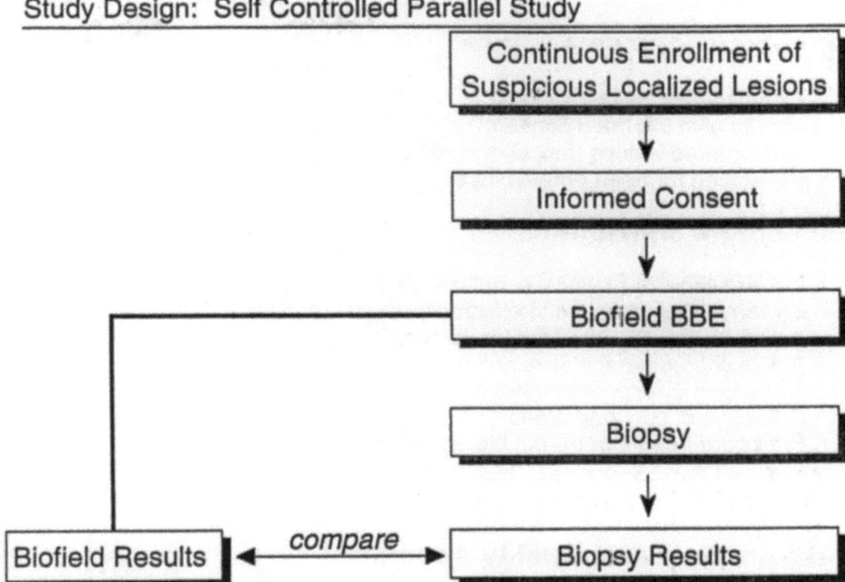

Fig. 2. This diagram shows the flow and quality assurance steps included in the self-controlled parallel study design

In the diagnosis of breast cancer, open biopsy with histopathological examination is the reference standard for the determination of malignancy. For this reason, the multicentre prospective trial incorporates open surgical biopsy as the reference endpoint. Patients with either image-detected or palpable breast masses scheduled for open biopsy will be eligible to participate in the trial. Prior to the open surgical biopsy, the patient will undergo a BBE using a directed sensor array.

In the masked study design (Fig. 2), the result of the BBE will not be available to influence clinical decision-making. After each examination, the electropotential differentials will be stored as an electronic data file which will be forwarded to an independent unblinding agent. The third party unblinding agent will process the recorded measurements using a predefined algorithm and compare the BBE result with the results of histopathological analysis derived from the patient's tissue biopsy.

This blinded design has several advantages in terms of eliminating possible sampling bias. In this type of clinical trial design, each patient serves as their own control. As the physicians are masked to the device output, there should be little possibility that the use of the BBE would have an effect on which patients are selected to go on to open tissue biopsy. In addition, the results of the BBE should not influence patient participation or the outcome variable

(i.e., the tissue diagnosis at open biopsy) as the pathologist will be masked to the result of the BBE.

In all clinical studies, it is important to be assured that subgroups of patients are not excluded from the study and that the inclusion criteria closely mirror the actual spectrum of clinical disease. Given that the BBE is completely non-invasive and does not involve additional exposure to radiation, electromagnetic fields, or any additional tissue or blood sampling, investigators in the initial pilot studies have found very high patient acceptance rates for the BBE. The inclusion requirements are generous and require only that the woman be capable of informed consent and that no potential confounding conditions are present (see inclusion and exclusion criteria, Table 1).

One of the strengths of the chosen study design is that, no matter what exact sequence of diagnostic evaluation is followed, a patient is eligible for the study provided she is scheduled for open biopsy. Thus across the centres involved in the study, patients who are eligible and enrolled represent a temporal cross-section of clinical practice and decision-making. Centres have been invited to participate in the study in order to achieve a demographic and geographic sample of patients with breast lesions from France, Germany, Great Britain, Italy, and the Netherlands.

Table 1. Inclusion and exclusion criteria for clinical studies with BBE

INCLUSION CRITERIA

- Females over 18 years of age
- Able to give informed consent
- One or more lesions (palpable or identified by imaging modalities)
- Scheduled for open breast biopsy

EXCLUSION CRITERIA

- Dermatologic or physical condition precluding the application of all necessary sensors
- Known allergic reaction to electroconductive medium (ECM) or adhesive material used to apply sensors
- Active electrically powered implanted devices
- Active chemotherapy or radiation therapy within last 3 months
- Tamoxifen therapy within last 3 months
- Presence of only one breast
- Prior core-needle or tru-cut biopsy of lesion to be tested

Data Analysis and Quality Assurance

The multicentre study design permits a high degree of data validation as the original BBE results are obtained and retained at each study site as primary source documentation, both as a paper hardcopy and as an archival electronic data file. Copies of the BBE measurements are sent to the unblinding agent (Imperial Cancer Research Fund, Breast Cancer Epidemiology Unit) which verifies the BBE readings and then compares the BBE result with the pathology report - a source document derived from the original surgical specimen.

Pathological examination involves the evaluation of tissue and cellular architecture and staining patterns as interpreted by an expert pathologist. The identification of invasive ductal carcinoma of the breast usually shows a high degree of reproducibility between expert pathologists. Published studies have suggested considerable between-observer variability exists for the diagnosis of ductal carcinoma *in situ* (DCIS) and atypical hyperplasia [15], with a much lower within-observer variation. As a result, all pathological slides included in the multicentre study will be reviewed by a referee pathologist to reduce interobserver variation. For the purposes of this study, DCIS will be considered malignant and atypical hyperplasia will be considered as a benign lesion.

Sensitivity and specificity will be calculated in the standard manner, where sensitivity is equal to the number of true positives divided by the number of true positives plus false negatives [5]. Specificity is derived from the number of patients with negative test results who are found to have a benign process at biopsy; more formally stated, specificity is equal to true negatives divided by true negatives plus false positives [5].

Given a patient with either a positive or negative test result, the most relevant question is: what is the likelihood that the patient actually has disease. The post-test probability that a patient with a positive test actually has the disease can be determined by the prevalence of the disease and the calculated sensitivity and specificity of the test. The exact conditional probability can be calculated using Bayes' rule. More commonly, the terms predictive value (positive) and predictive value (negative) are used to describe the post-test probability of disease [5,9].

It is of great interest that with the reported 97% sensitivity and 80% specificity of the BBE in the pilot clinical trials [this monograph and 18], the BBE has the potential to function as a powerful post-screening filter with an exceptionally high negative predictive value (NPV). In post-screening populations where the prevalence of breast cancer may range from a low of 1% to a high of 10%, this would yield an NPV of 99.6% to 97.8%.

Conclusion

The BBE has been shown to be a promising modality for the non-invasive detection of breast cancer in pilot clinical studies. These studies have evaluated a sample of the appropriate spectrum of disease necessary to create a construction set for further testing. The multicentre clinical evaluation employing a masked, self-controlled outcome design, using tissue biopsy as an endpoint, appears to be a robust prospective assessment of clinical utility.

The diagnostic BBE as employed in these trials is a non-biasing measurement as the performance of the test does not affect the lesion or the outcome variable. The BBE results cannot be influenced by user-selected parameters and the masked nature of the result protects the study from inadvertent patient selection bias. In the planned multicentre study, the BBE result cannot be used to influence the selection of which patients go to biopsy. Studies of the BBE across a variety of European practice settings should provide an excellent assessment of its clinical utility, independent of variation in local practice, referral patterns, and thresholds for clinical decision-making.

REFERENCES

1 Galen RS and Gambino SR: Beyond Normality: The Predictive Value and Efficiency of Medical Diagnoses. Wiley, New York 1975

2 Shapiro AR: The evaluation of clinical predictions. A method and initial application. N Engl J Med 1977 (296):1509-1514

3 Habbema JDF and Hilden J: The measurement of performance in probabilistic diagnosis. IV. Utility considerations in therapeutics and prognostics. Methods Inf Med 1981 (20):80-96

4 Greenhouse SW and Mantel N: The evaluation of diagnostic tests. Biometrics 1950 (6):399-412

5 Vecchio TJ: Predictive value of a single diagnostic test in unselected populations. N Engl J Med 1966 (274):1171-1173

6 Beck JR, Meier FA, Rawnsley HM: Mathematical approaches to the analysis of laboratory data. Prog Clin Pathol 1981 (8):67-100

7 Ransohoff DF and Feinstein AR: Problems of spectrum and bias in evaluating the efficacy of diagnostic tests. N Engl J Med 1978 (299):926-930

8 Wasson JH, Sox HC, Neff RK, Goldman L: Clinical prediction rules. Applications and methodological standards. N Engl J Med 1985 (313):793-799

9 Linnet K and Brandt E: Assessing diagnostic tests once an optimal cutoff point has been selected. Clin Chem 1986 (32):1341-1346

10 Spiegelhalter DJ: Probabilistic prediction in patient management and clinical trials. Stat Med 1986 (5): 421-433

11 Pauker SG and Kassirer JP: The threshold approach to clinical decision making. N Engl J Med 1980 (302):1109-1117

12 Day NE, Williams DR, Khaw KT: Breast cancer screening programmes: the development of a monitoring and evaluation system. Br J Cancer 1989 (59):954-958

13 Frisel J, Eklund G, Hellstrom L et al: Randomized study of mammography screening - preliminary report on mortality in the Stockholm trial. Breast Cancer Res Treat 1991 (18):49-56

14 Nystrom L, Rutqvist LE, Wall S et al: Breast cancer screening with mammography: overview of Swedish randomised trials. Lancet 1993 (341):973-978

15 Tubiana M, Holland R, Kopans DB, Kurtz JM, Petit JY, Rilke F, Sacchini V Tornberg S: Advisory report of the European School of Oncology to the Commission of the European Communities "Europe Against Cancer" Programme. Advisory Report 1993

16 Muir Gray JA: General principles of quality assurance in breast cancer screening. NHS breast screening programme. Department of Community Medicine, Radcliffe Infirmary, Oxford, UK

17 Schnitt SJ, Connolly JL, Tavassoli FA, Fechner RE, Kempson RL, Gelman R, Page DL: Interobserver reproducibility in the diagnosis of ductal proliferative breast lesions using standardized criteria. Am J Surg Path 1992 (16/12):1133-1143

18 Weiss BA, Ganepola APG, Freeman HP, Hsu Y-S, Faupel ML: Surface electrical potentials as a new modality in the diagnosis of breast lesions - a preliminary report. Breast Dis 1994 (7):91-98

If you have any concerns about our products,
you can contact us on
ProductSafety@springernature.com

In case Publisher is established outside the EU,
the EU authorized representative is:
Springer Nature Customer Service Center GmbH
Europaplatz 3, 69115 Heidelberg, Germany

Printed by Libri Plureos GmbH
in Hamburg, Germany